Elegant Anatomy

History of Science and Medicine Library

VOLUME 47

The titles published in this series are listed at *brill.com/hsml*

Elegant Anatomy

The Eighteenth-Century Leiden Anatomical Collections

By

Marieke M. A. Hendriksen

TUTA SUB AEGIDE PALLAS · 1683 ·

BRILL

LEIDEN | BOSTON

Cover illustration: LUMC Al0007. Heart injected with red wax and mercury. Probably from the collection of Eduard Sandifort (1742–1814). © Anatomisch Museum LUMC 2012. Photographer: Arno Massee.

Library of Congress Cataloging-in-Publication Data

Hendriksen, Marieke M. A.
Elegant anatomy : the eighteenth-century Leiden anatomical collections / by Marieke M.A. Hendriksen.
pages cm. — (History of science and medicine library, ISSN 1872-0684 ; volume 47)
Includes bibliographical references and index.
ISBN 978-90-04-26278-2 (hardback : acid-free paper) — ISBN 978-90-04-26277-5 (e-book) 1. Anatomical museums—Netherlands—Leiden—History—18th century. 2. Human anatomy—Netherlands—Leiden—History—18th century. 3. Human anatomy—Netherlands—Leiden—Methodology—History—18th century. 4. Human body—Social aspects—Netherlands—Leiden—History—18th century. 5. Material culture—Netherlands—Leiden—History—18th century. 6. Senses and sensation—Social aspects—Netherlands—Leiden—History—18th century. 7. Aesthetics—Social aspects—Netherlands—Leiden—History—18th century. 8. Art and science—Netherlands—Leiden—History—18th century. 9. Leiden (Netherlands)—Intellectual life—18th century. I. Title.
QM51.N42L34 2015
611.0074'49238—dc23

2014033549

This publication has been typeset in the multilingual 'Brill' typeface. With over 5,100 characters covering Latin, IPA, Greek, and Cyrillic, this typeface is especially suitable for use in the humanities.
For more information, please see brill.com/brill-typeface.

ISSN 1872-0684
ISBN 978-90-04-26278-2 (hardback)
ISBN 978-90-04-26277-5 (e-book)

Contents

Acknowledgements

This book was originally conceived as a PhD thesis between 2008 and 2012 within the research programme *Cultures of Collecting: The Leiden Anatomical Collections in Context* at Leiden University and funded by NWO (the Dutch Research Council). Rina Knoeff, Robert Zwijnenberg, and Hieke Huistra were the other researchers on the project, and their ongoing support has been invaluable. In 2011, Simon Chaplin, director of the London Wellcome Library, gave Hieke and myself the opportunity to do research in London for three months. This period, with access to the marvellous collections at the Wellcome Library, the Hunterian Museum and the British Library, has been essential for the development of the book. My then home institution, the Leiden University Centre for the Arts in Society (LUCAS, formerly LUICD), Stichting Historia Medicinae, the Leids Universiteits Fonds and the Scottish Society for Art History generously funded this research trip.

Feedback from participants at international workshops and conferences in London, Lyon, Leiden, Zeist, Durham, Edinburgh, Zürich, Paris, Warwick, Amsterdam and Manchester on parts of the research underlying the current book have helped me greatly in testing and developing my ideas. In Leiden, there was also the ongoing support of our informal reading group—Frans van Lunteren, Martin Weiss and Eric Jorink regularly joined our small research group to discuss work in progress. Also very important for the development of my work were the international project workshop and conference we organized in Leiden in 2009 and 2012 respectively. The latter has formed an international network of scholars, curators, and artists working with anatomical collections, and has resulted in the Leiden Declaration on Human Anatomy.

The feedback of the organizers, my fellow PhD candidates and Dick Willems (professor of Medical Ethics at the Amsterdam Medical Centre/University of Amsterdam) during the meetings of the national WTMC research school has helped me to see my work in a broader perspective. Numerous people at universities, museums, libraries, archives, and research institutions have helped me find sources and access collections over the years. It is impossible to name everyone here, but special thanks are due to Andries van Dam, curator at Leiden University Anatomical Museum, for putting up with my endless inquiries, helping me access remote storage spaces, explaining the technicalities of preserving anatomical preparations, tolerating my presence in his not-so-big office and his sense of humour. Evan Ragland provided valuable feedback on the finished thesis when I was in the process of rewriting it as a book.

Others who have helped me are Ernst Otto Onnasch and Paul Ziche (Utrecht University), Bas Wielaard (LUMC), the staff at the Leiden University Library Special Collections, the people at Museum Boerhaave, especially Mieneke te Hennepe, Bart Grob, Tim Huisman and the library staff, Margot Barteling (LUMC), Dave Mazierski (University of Toronto), Marcel van der Beek and Ans ter Woerds at Geldmuseum Utrecht, Huib Zuidervaart (Huygens ING), Laurens de Rooy (Museum Vrolik), Annette Schmidt (Museum Volkenkunde, Leiden), Ingeborg Eggink, Wayne Modest (KIT/Royal Institute for the Tropics), Ricky Tax (Museum Meermanno-Westerianum, The Hague), Stephen Snelders (UMCU), and Reina de Raat, Jan Willem Pette and Paul Lambers at Universiteitsmuseum Utrecht. Without the technical knowledge and skills of Tiemen Cocquyt, curator at Museum Boerhaave, experimenting with making a preparation would not have been possible.

Then there is all the helpful staff at the Utrecht and Amsterdam University Libraries, the Royal Library and National Archive in The Hague, the City Archives in Leiden and Middelburg, the London Linnean Society and the curators that showed their collections and answered questions: Samuel Alberti at the London Hunterian Museum, Andrew Cornell at the Royal College of Surgeons of Edinburgh, Martha Fleming at the London Natural History Museum, and Maggie Reilly, curator of zoology at the University of Glasgow. In addition, I thank Teunis van Heiningen for helping me access some of Brugman's letters, and professor Marco de Ruiter for letting Hieke and me attend anatomy classes at the LUMC. The Museum Committee at the Leiden University Anatomical Museum should be credited for their overall critical support of this project. Leanne Jansen and Eva van Weenen have checked and helped me improve my somewhat dodgy translations and interpretations of sources in Latin.

Most recently, my colleagues at the University of Groningen have supported me in the complicated process of turning a thesis into a book. The help, suggestions, and critique of all mentioned here have been invaluable, yet all mistakes are obviously my own.

Utrecht, November 2013

List of Illustrations

CHAPTER 1

Introduction: Understanding Eighteenth-Century Leiden Anatomy

On a rainy cold January afternoon in 2012, eight people gathered in an attick workshop in Museum Boerhaave in Leiden, the Netherlands. I put my plastic bag on the table and unwrapped two sheep hearts and a liver, while Hieke laid out photocopies from an eighteenth-century anatomical handbook, Tiemen heated mixtures of wax, resin, colourants and turpentine on hot plates, and Andries heated an au-bain-marie and laid out heavy glass and metal syringes and medical scalpels. After distinguishing the pulmonary veins from the inferior vena cava, and the aorta from the pulmonary artery, which in itself is quite difficult for people without medical training, we started the process of injecting one of the hearts with the coloured wax according to the eighteenth-century instructions. After two hours of trying to find the ideal temperature for the wax mixtures, syringes, and bain-marie—which had to keep the heart warm enough for the wax not to solidfy immediately—we ended up light-headed from the turpentine fumes, and with a coarsely injected and slightly cooked heart. I was in the final year of my PhD research on eighteenth-century anatomical collections, and finally fully understood the curiosity, the immense craftsmanship and the endless patience embodied in the preparations that I had studied for over three years.

It was exactly in order to enhance my own understanding of the material qualities of anatomical preparations and the physical experience of creating them that I attempted to reproduce the process of making an anatomical preparation during two sessions in early 2012. With the help of fellow *Cultures of Collecting* project members Rina Knoeff, Hieke Huistra, technical support from Andries van Dam, curator of the anatomical museum at the LUMC, and Museum Boerhaave curator Tiemen Cockuyt, I tried to follow a 1790 recipe by Thomas Pole for creating wax-injected preparations of a (sheep) heart and a liver.[1] This experiment was inspired by William R. Newman's recreations of Newton's alchemical experiments, which can be viewed online, and by Pamela Smith's work on the sixteenth-century craftsman and metal caster Jamnitzer.[2] Smith, together with staff and students at the University of Amsterdam, tried

1 Pole, *The anatomical instructor*, xiv, xxiv–xxv.

2 Newman, William R. *The Chymistry of Isaac Newton*, 2005–2014. http://webapp1.dlib.indiana.edu/newton/reference/chemLab.do.

 | DOI 10.1163/9789004262775_002

to create casts of animals and insects based on instructions by Jamnitzer. She argues that although exploring historical objects through reconstructing them may seem overly emphatic, it may work as you have an external object to compare your reconstruction to. It is hard to capture the experience of reconstruction in writing, but it can be an efficient way of gaining knowledge of techniques and insights into historical accounts. Smith stressed that after trying to do something yourself, reading a historical book or manuscript on that technique is no longer a linear experience; it becomes more holistic.[3] This is defintely true, as I discovered when attempting to recreate an anatomical preparation.

As reconstruction is not an option for everyone, this book, with its close readings of a select number of the hundreds of eighteenth-century anatomical prepararations housed in the Leiden University Medical Center's anatomical museum, in Museum Boerhaave, the Leiden-based Dutch museum for the history of science and medicine, and in various storages around Leiden, aims to provide an exploration of the experiences, material and epistemic cultures from which these historical objects result. This book is not intended to give a linear or chronological, institutional account of eighteenth-century Leiden anatomy and its leading luminaries. Although there is not a single concise overview of the kind, these topics have been sufficiently described in Leiden University histories such as those by Molhuysen and Otterspeer, in Barge's early twentieth-century histories of early Leiden medical teaching and collections, in Huisman's work on the anatomical theatre in the seventeenth century, and in several biographical accounts of Leiden anatomists, such as Punt's work on B.S. Albinus, Zwaag on Van Doeveren, Van Heiningen on Gerard Sandifort and the correspondence of S.J. Brugmans, and of course Luyendijk-Elshout's (nee Elshout) work on the eighteenth-century Leiden anatomical cabinet and Eduard Sandifort.[4]

The existing literature tells us many things about lives, discoveries and careers, and also sometimes about university politics and personal relations. The aforementioned works have great value, and form a sturdy basis for my present purposes. However, they tend to tell us very little about what we now call the eighteenth-century Leiden anatomical collections, and even less about

---

3 Smith, *The workshop of history*, Smith & Beentjes, "Nature and art."

4 Molhuysen, *Bronnen*, Otterspeer, *De vesting van macht, De werken van wetenschap*, Barge, *De Leidsche anatomie herdacht, Het geneeskundig onderwijs, De oudste inventaris, De stiching*, Punt, *Bernard Siegfried Albinus*, Van der Zwaag, *Wouter van Doeveren*, Van Heiningen, *Gerard Sandifort, The correspondence*, Elshout, *Het Leidse kabinet*, Luyendijk-Elshout, "Antony Nuck," "The elegant anatomist," Huisman, *De volmaakte mens, The finger of God.*

the specifics of individual preparations. As the actual merging of eighteenth-century anatomical preparations into a collection in Leiden was a process that took place roughly between 1770 and 1820 they are usually not understood as a whole. Moreover, the object-driven approach is a fairly recent one in the history of science. Therefore the biographies and institutional histories currently available shed little light on the eighteenth-century collections as a whole or on particular eighteenth-century Leiden anatomical preparations and the specific circumstances in which they were made and used, or on how they were looked at, both directly after their making and upon acquisition by the university. This book aims to fill that gap, and thus to provide a new understanding of eighteenth-century Leiden anatomy.

As most of the existing works on Leiden anatomy do not have the preparations or collections as a starting point, the story they tell about them is fragmented and diffused. The few works that do take the Leiden anatomical collections as a starting point are dated and rather brief, such as Geyskens' and Van der Klaauw's 1934 report.[5] Elshout's description of the collections from 1952 was primarily a conservation report with added biographies of the anatomists who made the preparations, but she was the first to actually devote some attention to the eighteenth-century anatomical preparations as cultural-historical objects.[6] That approach is important, as what we are left with today are the preparations themselves: delicate, highly evocative, though foreign objects, which hint at a material and academic culture entirely different from the one we know. In this light, maybe the most remarkable thing is that almost none of the existing literature raises why questions about the eighteenth-century Leiden anatomical collections: why do they look the way they do and why is there so much variation in their subjects, in the techniques used to make them and in their finishing? Why do they evoke both appreciation and disgust? Why were some used in teaching for decades and heralded as tokens of the greatness of Leiden University, while others, although equally skilfully made, barely saw the light of day and were persistently ignored after their formation? Why did these preparations end up in this collection while other, apparently equally suitable, candidates did not? Answering these questions matters, as the answers clarify their significance for the history of science, for cultural history and for the history of collections and collecting. In addition, they may help to inform decisions about preservation and display.

Other academic works on particular historical anatomical collections have appeared recently, such as Simon Chaplin's *John Hunter and the 'museum*

---

5 Geyskens and Van der Klaauw, "Der heutige Zustand."

6 Elshout, *Het Leidse kabinet*.

oeconomy,' 1750–1800 (2009), Anna Maerker's *Model Experts* (2011), and Rebecca Messbarger's *The Lady Anatomist* (2010). This book distinguishes itself from these works by covering a Dutch collection that has not been studied integrally before and by using the objects in the collections as its starting point, instead of the person(s) that created them. The actual, physical, material objects are the starting point for this book, as most of the why-questions are evoked by the materiality of the preparations. This means that I will borrow freely from methods and approaches originating in the fields of philosophy, material culture studies and art history.[7] In short, this book is object-centred; it uses precise descriptions, focussing on the internal evidence of the object, followed by deduction and speculation, exploring possible connections between objects, people, and their intellectual and emotional responses to the objects, drawing from a variety of sources.[8] This will eventually lead to a better understanding of why the object is the way it is, what its historical value is, and why it provoked certain responses from its makers and contemporary users.[9]

As Alberti has pointed out, museum objects gather meanings through association with the people that make and use them, thus linking their history and that of the collections of which they are a part to broader scientific and civic cultures.[10] The objects themselves, their materiality, possess some kind of agency in this process, although the agency of objects appears to be of a different nature than that of human actors.[11] In this book, the main focus will there-

7 Obviously I am not the first historian of science doing this. Notable examples are Lucia Dacome, Lorraine Daston, Ursula Klein and Pamela Smith.

8 The concept of internal evidence is derived from Prown, "Mind in matter," 9, *Art as Evidence*, 235–242. By internal evidence, Prown means any facts about an object that can be derived from our sensory perceptions of its materiality, preceded or accompanied by intellectual apprehension.

9 For more on object-centred material culture studies see i.e. Harvey, "Introduction: Practical Matters," in *History and material culture*, Herman, "The Discourse of Objects," and Prown, *Art as Evidence*. Also see the object-oriented ontology in literary theory, i.e. Harman, "Tool-being," and Morton, *Ecology Without Nature*.

10 Alberti, "Objects and the museum," 599.

11 Hacking, *Historical Ontology*, 17, Latour, *We have never been modern*, 142–145, *Reassembling the social*, 63–87. Object-oriented ontology (OOO), and to a lesser extent Latour's Actor Network Theory, put things at the centre of the study of existence. OOO's most radical proponents contend that nothing has special status; everything, whether human or object or phenomenon, exists equally. My idea of the agency of objects is similar to what Hacking describes as the agency of nature: he does not, like Latour, advocate a parliament of things or quest to minimize the difference between the human and the non-human. Hacking states that nature has some part in the development of beliefs about it, but that it has limited agency.

fore be on the materiality of the anatomical preparations, their manufacture and collecting, and to a lesser extent their use within the collections, as it is through their materiality that the eighteenth-century Leiden anatomical preparations have played and will play a part in creating and shaping experiences, identities, and relationships.[12] That materiality was most obvious to the people who created and collected them, as they experienced it first-hand, whereas for later users of the collections this was largely (but not entirely) reduced to the visual realm. The materiality of the preparations is, in short, my primary point of access to their meaning.[13]

As our own experiments clearly showed, the circumstances of an eighteenth-century dissection room cannot be reproduced, as the moral and hygienic objections would simply be too big. However, even our relatively simple reproduction experiment brought home truths about eighteenth-century practices of making anatomical preparations more emphatically than reading any number of sources and studying any number of historical preparations ever could have, even if the results were a far cry from the refined, lasting preparations discussed in this book.[14] First, the sheer amount of work involved in the preparations and actual creation of our 'preparations' was striking. It took two people several weeks to get together the equipment and materials needed to start the experiment, and things like syringes were rather harder to obtain in the eighteenth century then they are now—they would have had to be made to order. Moreover, we simply obtained our 'specimens' from a butcher, but even for an anatomy professor, it would have often been difficult to obtain suitable specimens for preparations—in fact, early modern anatomy professors also sometimes relied on butchers or slaughterhouses for animal specimens if human corpses were not available.[15]

---

12 Alberti, "Objects and the museum," 561. Alberti distinguishes three phases in the life of a museum object, of which the manufacture and collecting is the first and the use of the object within a collection is the second. The third phase, the role of the object in the experience of the museum visitor, will only be taken into account here in as far as it is accessible and relevant for the understanding of the contemporary, eighteenth-century biography of the object.

13 Pickstone, *Ways of knowing*, 216. Pickstone reminds us of the obvious but frequently ignored fact that our primary relationship to nature, as to each other, is one of meaning. I would like to add our primary relationship to objects to that.

14 For a more detailed account of this reproduction experiment, see Hendriksen, "Aesthesis in anatomy," Appendix II.

15 Richardson, *Death, dissection*. Ruth Richardson has vividly discussed the excesses to which the difficulties in obtaining bodies for dissection could lead.

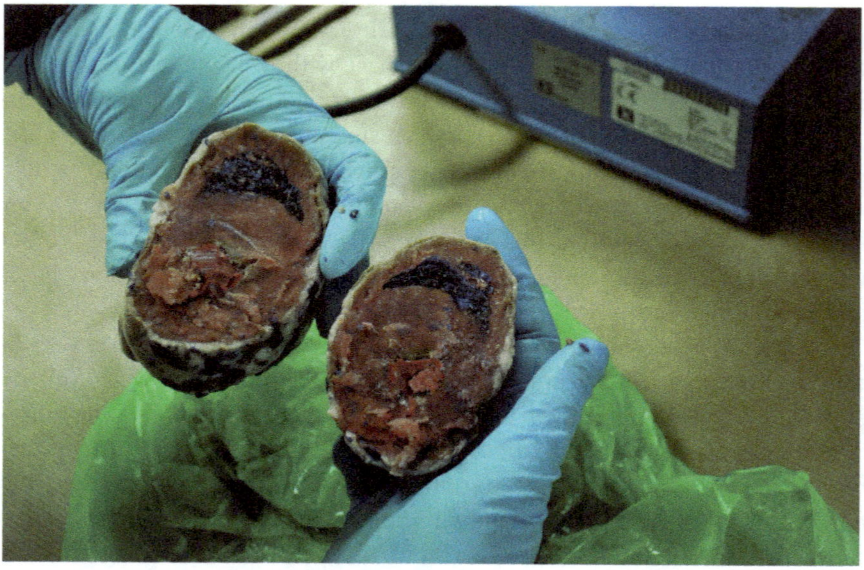

ILLUSTRATION 1.0 *Sheep heart, cut trough horizontally, a week after coloured wax had been injected in early 2012. Although red and blue wax is clearly visible in the heart chambers, the wax has not entered any of the smaller vessels and was not visible from the outside. Photograph: author.*

Second, even if the natural philosophers that created the eighteenth-century preparations outsourced some of the shopping, creating preparations remains a labour-intensive and hard to fathom process. Our experience affirmed that the task of making anatomical preparations relies largely on tacit knowledge. No handbook in the world can explain the exactitudes of the practice. For example, how coarse or thin the injection mass should be depends on numerous factors, like the structure of the organ one chooses to inject, the desired penetration of vessels, and the temperature of the tissue, the injection mass, the room as well as the syringe. The size of both the organ to be injected and that of the syringe and needle matter greatly, as well as how many people there are around to assist. Learning how to make a well-injected preparation is an art that can only be learned through endless practice, through trial and error, with hands-on work. It is truly about gaining knowledge from all your sensory perceptions—we used our touch, smell, sight and hearing, and it takes little imagination to envision the inclusion of taste.

Because of this necessary sensory involvement, making an anatomical preparation also means dealing with disgust in different degrees. The initial queasiness we felt when unwrapping the raw sheep hearts waned quickly; we got used to the smell and feel, and they were nowhere near a discernible state of putrefaction yet, so there was little about them that really bothered us. The liver really did smell bad, but that was quickly forgotten once we got impressive results injecting it. The turpentine fumes coming off the injection mass were not exactly disgusting, but they were smelly and made us light-headed. Of course these organs were bought from a butcher, who had 'dissected' them from a freshly slaughtered animal body and cleaned them, withholding from us the visceral experiences of having to harvest organs from a not-so-fresh subject ourselves, so our disgust must have been minimal compared to what could be experienced in an eighteenth-century dissection and preparation room.

However, despite the initial disgust, fascination quickly took over. What had looked so straightforward in the pictures in the anatomical atlas, and seemed so uncomplicated in Pole's description, was much more complex in practice. Even working out what vein to inject turned out to be a problem. Yet these complexities also ignited a strange kind of admiration for nature—its variety, and the fact that the sheep hearts were all slightly different. Yet presumably they had all worked absolutely fine, they had kept alive fairly big mammals, and in a way they were all beautiful and perfect. This excitement and admiration intensified with the injection of a liver. Before the injection, it was a stinky, slippery clump of tissue, and in the anatomical atlas it looked like a wiry piece of machinery of some kind. But when the blue wax was injected, all of a sudden wonderfully refined vein structures became visible. These experiences made it easier to understand why the eighteenth-century Leiden anatomists wanted to reveal and exhibit the beauty and perfection they found in their anatomical practices in refined preparations.

What we also experienced distinctly was the fact that materials are resistant. The sheep hearts in particular resisted injection, the wax masses resisted remaining fluid long enough to penetrate the vessels we wanted them to, and even after putting the preparations in jars, they resisted remaining stable: the coloured wax dissolved in the alcohol, slowly undoing our work and rendering the preparation invisible. Even though we did not intend to sell them or put them on display, we nevertheless found that the materiality of things we used to try to make elegantly injected preparations resist commodification, or at least objectification. In fact, the aspects of the aesthesis that shaped the eighteenth-century anatomical preparations and the epistemic culture they were a part of, were still to some extent discernible in our attempt to reproduce an eighteenth-century wax-injected preparation.

However enlightening the experience of attempting to make an anatomical preparation is, it would be highly impractical to try to make visitors of historical anatomical collections aware of the various aspects of aesthesis through making them reproduce preparation techniques. Yet we do want to provide ways for them to connect to historical objects such as these preparations. As Legêne points out, we therefore need to provide disclosure about the history of objects, the context in which they were collected and the aim of their presentation, because otherwise visitors are unlikely to identify with them and establish a connection with a shared, if contested, history.[16] Historical anatomical collections matter because they are evocative reminders of our fairly recent history of medicine, which is in fact a cultural history.

The Leiden collections have so much to tell us about how our forebears thought about health and sickness, life and death, beauty and ugliness, research and medicine, themselves and the other, and how their ideas and practices influenced ours, that it would be a loss if we did not try to enhance our understanding of them. This book therefore aims to explore the eighteenth-century Leiden anatomical collections to gain a better understanding of why and how these objects were made and used, of what they meant and can mean.

The starting point for these explorations is the very specific materiality of the preparations. To do so systematically, and to understand the epistemic culture from which the preparations emerged, a theoretical framework is used to develop the analytical concept of aesthesis in the first chapter. Aesthesis has a number of tacit yet distinct, interwoven characteristics, such as gaining knowledge through sensory perceptions, a search for beauty and perfection in those perceptions, dealing with disgust evoked by the practices of anatomy, and commodification of the body. The subsequent chapters each focus on a number of particular preparations that share some kind of material characteristic, i.e. the presence of mercury, lace, plants or bead strings, or pathology and deformity, or simply consisting of the same bodily tissue, like bone. From the analyses of their materiality, it will appear that the preparations and the epistemic culture from which they resulted can be understood in new ways through the concept of aesthesis and its various aspects. Although all aspects can be found in every preparation to some extent, in most one particular aspect stands out distinctly. The final chapter on bone preparations explores the end of aesthesis.

As for the selection of preparations discussed in this book, it is far from exhaustive. Only twenty-nine of the 8170 preparations believed to be from the eighteenth century are explicitly described and depicted here. Of those twenty-nine, only six are dried preparations, whereas almost seventy per cent of the entire eighteenth-century collection consists of dried preparations. This may seem disproportionate; an overrepresentation of wet preparations. However, the wet preparations shown here, as well as the dried preparations in

16 Legêne, *Flatirons and the folds of history*, 58.

the selection, are what is particular for Leiden anatomy in this period. Before 1700, the creation of lasting wet anatomical preparations was very limited and mostly unsuccessful. In the course of the eighteenth century, as will also be discussed in more detail in chapters two and three, preparations techniques and preparation fluids became increasingly refined, and Leiden was one of the main European centres of these developments. Dried preparations existed before and after the eighteenth century too, and although wet preparations are still made today, the refinement and particular iconography of eighteenth-century wet preparations has hardly been equalled before or since.

Moreover, the preparations discussed in this book are representatives for particular kinds of eighteenth-century preparations that occur in other collections from the period too, both in the Netherlands and abroad. For exampe, mercury-injected preparations can also be found in the contemporary collections of John and William Hunter in London and Edinburgh, and in eighteenth-century preparations in Halle (Germany). Additions of beads and plants are also seen in specimens preserved by anatomists in Amsterdam and Groningen. The objects chosen therefore form a representative and appealing selection of the Leiden collections and are indeed a unique access point to an understanding of their history.

CHAPTER 2

Elegant Anatomy: Aesthesis

This book argues that the answer to many of the why-questions evoked by the materiality of the eighteenth-century Leiden anatomical preparations—most importantly, why do they look so distinctly different than modern anatomical preparations?—lies in aesthesis. Aesthesis is a term that occurs in eighteenth-century texts and dictionaries, albeit rather sparingly. It is not usually found in general and art dictionaries before the nineteenth century, but two examples appear in medical and philosophical dictionaries. In Barrow's 1749 *Dictionarium Medicum Universale: Or, a New Medicinal Dictionary*, 'aesthesis' is explained as "to be sensible of, sensation, or the faculty or power of sensation." The same definition is given in James's 1740s medical dictionary. These appearances indicate that aesthesis was deemed specifically relevant in a medical context in the eighteenth century. Therefore aesthesis is used as an analytic category here, in search for a concept that can define a particular epistemic culture. It is very useful as it is a contemporary, eighteenth-century term denoting a reliance on the senses.[1]

Aesthesis referred to a certain sensibility, the sensory perceptions of materiality, yet the word is also closely related to the terms 'aesthetica' and 'aesthetics', a connection that will be discussed in more detail in this chapter. This relation suggests aesthesis also had something to do with a sensibility to beauty and elegance in the sensory perceptions and materiality of anatomy, as well as possibly dealing with the disgust these may evoke too. Another aspect of aesthesis is commodification, the stabilizing and stressing of certain qualities of human tissue to make it into a tradable object. Yet it is the sensibility to beauty and perfection in materiality that is defining for aesthesis; it is what

1 Meyer, *Religious Sensations*. Pinney, "Four Types of Visual Culture," 131–45. The term aesthesis has been revived in a number of contexts recently. Anthropological work by Meyer and Pinney on corporeal and emotional responses to images by observers uses the notion of aisthesis in the "earlier, pre-Kantian sense of aesthetics denoting our total sensory experience of the world and our sensitive knowledge of it." Lehmann and Roodenburg. "Introduction." *Netherlands Yearbook for Art History*. Eds. Lehmann, Ann-Sophie and Herman Roodenburg. (Zwolle: Waanders Publishers, 2008), 7–13, 10. Johanna Drucker uses the term aesthesis to refer to 'a theory of partial, situated, and subjective knowledge' in the context of digital humanities. Drucker, *SpecLab: digital aesthetics*, xiii, chapter 3. However, unlike my use of the term, these revivals are unconnected to the original, medical context of aesthesis.

 | DOI 10.1163/9789004262775_003

distinguishes it from other sensuous epistemic cultures and commodification practices. The challenge is that most of its intrinsic qualities are often tacit rather than explicit. However, the interrelated elements of aesthesis transcend historiographies such as biographies, institutional histories and colonial history, and can therefore provide us with a new understanding of the anatomical preparations in eighteenth-century Leiden. This book is therefore an exploration of the characteristics of aesthesis in Leiden anatomy of that period.

The most problematic aspect of studying an epistemic culture is that its characteristics are hardly ever explicit—they can be known only through an analysis, both visual and material, of an anatomical preparation. Like any epistemic culture, the culture understood through aesthesis is tacit, and does not so much reside in an individual attitude as being a constitutive property of an epistemic culture. It could be said to be weak and relational, as described by Collins: it is not tacit because of deliberate secrecy, it is just so embedded, so obvious to its members, that there is no need for them to make it explicit. It does not stand out until seen through foreign eyes, and to those eyes, once they are used to this exotic country that is the past, it is clearly visible and all-encompassing.[2] Explicit references to the importance of sensory perception and the search for beauty therein are rather rare in the work of eighteenth-century Leiden anatomists. Particularly rarely mentioned is the quest to find and bring out beauty, perfection, and elegance in anatomy, and the use in any constructive way of the disgust it regularly evokes.

Aesthesis is primarily expressed in relation to material objects rather than being explicitly explained in texts. Although none of the people discussed in this book seem to have used the term aesthesis, the anatomical preparations they made and occasionally the written sources they left us show the importance of sensory perception in creating and studying anatomical preparations, and how it was combined with a search for beauty and perfection. Neither exactly an actor's category nor a neologism, my use of aesthesis as an analytical category therefore draws on contemporary definitions of the word and similar terms like aesthetics. This chapter will discuss more extensively what is meant by the aesthesis of anatomy, to avoid any possible confusion as to why I use aesthesis and not the related word aesthetics. We will also look at why aesthesis is essential for an understanding of eighteenth-century Leiden anatomy, and in what consists its added value for the history of science.

---

2 Collins, *Tacit and Explicit Knowledge*, ch. 4–6.

What is Aesthesis?

What exactly is meant by aesthesis in this book? Aesthesis has five important, interlinked characteristics. First, it is a concept that describes an epistemic culture, a manner of gaining knowledge rooted in sensory perception and materiality: it refers to a science of things that can only be known through sensory perception. Secondly, it scrutinizes and tries to manipulate the beauty, perfection and elegance, encountered in these sensory perceptions of the material world. Thirdly, it can also be a way of dealing with the disgust evoked by practicing anatomy. Fourthly, aesthesis in anatomy is inevitably characterized by the very tactile processes of commodification, domestication and objectification: it involves the creation of lasting, transferable anatomical preparations that both represent and are made of parts of the human body, as well as the domestication of the (exotic) other.[3] Finally, aesthesis is largely tacit. I will now discuss these characteristics in more detail.

Sensory Perception and Materiality

The first important characteristic of aesthesis is that it concerns the knowledge of things that can only be known through sensory perception. It is not surprising that it is in specifically eighteenth-century anatomy that this is so important. Ever since Vesalius took the scalpel in his own hand in the sixteenth century, anatomy had increasingly become a hands-on academic discipline, closely linked with the more philosophical discipline of physiology, which theorized anatomical findings.[4] Nonetheless, anatomy itself, laying bare and learning about the inner structures of the human body, was best done through the use of one's own hands and eyes. For example, in 1772, the newly appointed Leiden professor of anatomy, Eduard Sandifort (1742–1814), held his inaugural address in Leiden. In it, he stated that practical medical knowledge of both healthy and diseased anatomy, obtained through dissection, is essential for becoming a better physician and surgeon, as is also making a proper diagnosis in those cases where cutting does not occur.[5] To Leiden-trained anatomist

3 Latour, *Science in Action*, 223.

4 Cunningham, "The Pen and the Sword. Part One. Part Two."

5 Sandifort, *Oratio De Circumspecto Cadaverum*, 6: "Haec practicae medicinae Gubernatrix. Hac ad intimosabditosque organorum recessus paratur via, funestae mroborum exponuntur clades, tutior patet curandi methodus, & accuratior instituitur prognosis, ubi fatorum mandata rescindere non licet."

Petrus Camper (1722–1789), who would later become professor of anatomy at Groningen University, sensory perceptions were also of fundamental importance to all his work. Early in his career, Camper already stressed the importance of sensory perception in his dissertation on sight:

> God has wanted sensory perceptions, the testimonies of others, and reason by analogy, applied with the appropriate cautiousness, to lead us to the truth and to be the foundation of our beliefs.[6]

However, in an age without artificial refrigeration and before the advent of bacteriology, the visceral experience of practicing anatomy was not for the squeamish. Anatomical demonstrations were often spread over several days, and although being limited to the winter months, using an unheated room facing north, burning incense and starting with the most perishable parts—the intestines—went some way in covering up the worst in terms of smell, by the last day of dissecting the atmosphere in the anatomy theatre cannot have been pleasant.[7] This is not to mention the private dissections and the making of preparations many professors and, later, also students of anatomy were frequently involved in. It may seem that touch and sight would provide the only sensory perceptions needed to rely on for the anatomist. However, although the hand and the eye were indeed considered to be the anatomists' most important instruments, all senses, including smell, taste, and hearing were involved in the practice of anatomy. Wendy Moore has vividly described the sensory practices of the English eighteenth-century anatomist John Hunter, who actively encouraged his students to smell and even taste bodily fluids such as urine and semen in order to come to a post-mortem diagnosis.[8]

As appears from these above examples, aesthesis was characteristic of a sensory manner of gaining knowledge about the world, and they confirm what Pickstone has pointed out, namely, that all ways of knowing involve not merely mental operations but work as well: hands on, manual, practical work,

---

6 Camper, *Dissertatio optica de visu*: "Sensus, Testimonia, & Analogiam, adhibitis legitimis cautelis, nos ad veritatem conducere, & persuasionis fundamenta esse Deus voluit."

7 Cunningham, *The Anatomist Anatomis'd*, 26, 29–30, Huisman, *The Finger of God*, 31–32. These were only general guide lines for anatomical theatres, and it was not possible to follow them everywhere. In Leiden for example, the anatomical theatre was housed in an old church building, which meant there was a lot of natural light, but the room was facing east instead of north.

8 Moore, *The Knife Man*, 79–80.

such as experimenting and making preparations.[9] This I have myself experienced first-hand when trying to make a preparation. For example, based on the preparations I had seen and the preparation handbooks I had read, I expected the process of injecting internal organs with fluid coloured wax to be difficult because of the size of the veins, but I had not anticipated how difficult it would be to keep the wax hot enough during the process to remain fluid and injectable without burning my hands, nor had I expected the smell of turpentine to be so overwhelming. It is this kind of practical work that provides knowledge that would remain inaccessible when one only uses written sources.

In eighteenth-century aesthesis, collecting, mapping, and verifying sensory perceptions in anatomical preparations are precisely that work. When Leiden anatomy professor Wouter van Doeveren (1730–1783) obtained a preparation of a gravid uterus made by his deceased colleague Noortwijk for his own collections, he wrote an excited letter to a friend: finally he would be able to see and take apart the preparation, on which Noortwijk had based his thesis, that the blood circulation of mother and foetus are one, something Van Doeveren believed to be untrue based on his own experiments with gravid uteri. However, he apparently still felt it was important to verify and compare his own sensory perceptions and experiments with those of a colleague.

The importance of sensory perceptions and materiality was obviously not limited to the field of anatomy in the eighteenth century. The characteristics of the epistemic culture understood through aesthesis could only develop in a wider network, and the anatomical practitioners influenced by it were often equally involved in other fields of knowledge, such as natural history, medicine, chemistry, and fine arts. It is therefore very well possible that other collections and areas of knowledge from the same period can also be understood through aesthesis, a surmise supported by Lissa Roberts' 2005 chapter on 'The Death of the Sensuous Chemist'. Here she describes how the use of chemists' senses in the laboratory in the latter half of the eighteenth century was altered and constrained to various extents by the increased use of precision instru-

9 Pickstone, *Ways of Knowing*, 17–20. This is not to say that aesthesis is a way of knowing itself. Aesthesis can be found in several different ways of knowing as distinguished by Pickstone, such as hermeneutics (world-readings), natural history, analysis and experimentalism. Although Pickstone's ways of knowing form a valuable epistemological framework, they are not sufficient to understand the epistemic culture of which the eighteenth-century Leiden anatomical collections are the material remains. The concept of aesthesis, with its duality of sensory experience and seeking beauty and perfection, is the key to understanding that culture.

ments and mathematical calculation.[10] As we will see in this book, chemistry and anatomy were closely connected fields in the eighteenth century, and the use of the anatomist's senses also changed in this period, albeit for slightly different reasons.

The creation of lasting anatomical preparations, as well as the re-examination preparations made by others, was of course a way to map and disseminate knowledge, but the initial stages of dissecting a body and making preparations already had an independent epistemic value. It can seem that the resulting anatomical preparations have only visual aspects. The materiality of the preparations may indeed have been primarily a multi-sensorial affair for their makers, yet we will see that in many cases the eighteenth-century meanings of that materiality far exceeded the purely iconographic and visual for the viewer too. Moreover, although the importance of sensory experience in gaining knowledge in early modern and modern times has been stressed by other scholars in the past decades, what seems to distinguish aesthesis from this sensory sensibility is an added, inherent sensibility for elegance and perfection.[11] How, then, did aesthesis motivate the Leiden anatomists towards scrutinizing and attempting to manipulate the beauty, perfection, elegance, and disgust encountered in these sensory perceptions of the material body?

Beauty, Perfection, Elegance

Initially, it might seem strange that in an epistemic culture so dominated by sensory perception and the visceral, material body of both the anatomist and the anatomized subject, there is room at all for considerations of beauty, elegance and perfection. Yet both historical and contemporary responses to anatomical collections make clear that aiming for elegance and perfection played an important role in creating these preparations. Peter the Great was so impressed with the beauty of the body of a boy that the Amsterdam anatomist Fredrik Ruysch (1638–1731) had preserved as a whole that he kissed it when visiting Ruysch's cabinet in 1697, and on his second visit purchased the entire

10 Roberts, "Death of the Sensuous Chemist," 117, 123.

11 The work of Pamela Smith is a wonderful example of this. Other works that have contributed greatly to the understanding of the role of sensibility and sensory perception in epistemic cultures are, amongst others, Alpers, *The Art of Describing*, Gaukroger, *The Collapse of Mechanism*, Shapin & Schaffer, *Leviathan and the Air-Pump*, Riskin, *Science in the Age of Sensibility*, Smith, *The Body of the Artisan*, Roberts, "Death of the Sensuous Chemist," Roberts et al., *The Mindful Hand*, Schickore, *The Microscope and the Eye*.

collection of over 2,000 preparations and shipped it to Russia. Contemporary examples of the appreciation of the refinement and eerie beauty of historical anatomical preparations can be found in coffee-table books and blogs devoted to them.[12]

When we take a closer look at what eighteenth-century anatomists said and did concerning the reaching for beauty, perfection, and elegance in their work, it quickly becomes clear that this striving was closely connected to the sensory knowledge and materiality of anatomical practice. For example, the aforementioned Camper explicitly linked the importance of sensory perceptions to his quest for beauty when writing about endlessly measuring, and even cutting in half, the heads of deceased people of different ages and races: "I have tried to derive from the foundations of Nature itself the characteristics of true beauty in our heads."[13] His teacher, the famous Leiden professor of anatomy Bernard Siegfried Albinus (1697–1770) opened his 1736 dissertation on the arteries and veins of the human intestines with the words "First I will deal with the arteries and veins in man's intestines, now that I have obtained the opportunity to show these matters with exceptional skill and elegance (...)."[14] Here we read, on the one hand, that the exceptional, manual skills of the anatomist—his craftsmanship—is needed in order to know, and make perceivable, subtle anatomy, while on the other hand he calls his craftsmanship elegant. The concept of aesthesis helps to define and understand this combination of reliance on sensory perceptions and the pursuit of beauty and perfection—because here that is precisely what elegance means.

Eighteenth-century Leiden anatomy has previously been described as subtle and 'elegant' by Luyendijk-Elshout in the twentieth-century meaning of the word: appearing pleasantly because of its refined design.[15] Yet we should not forget that elegance in eighteenth-century Latin ('elegantia') has a double meaning: it refers both to refined taste and to precision and perfection. So in generating tactile and visual perceptions in dissection, acting with, and searching for, a certain kind of refined beauty and perfection is necessary

12 Gould and Purcell, *Finders, Keepers: Eight Collectors*, Fuchs and Fuchs, *Conserving*, De Rooy and Van den Boogaard, *Verzamelaars Van Vorm*, Ebenstein, *Morbid Anatomy* (blog), http://morbidanatomy.blogspot.com/.

13 Camper, *Verhandeling*, 96: "Ik heb getracht uit de gronden der Natuur zelve de kenmerken van het waare schoone in de hoofden van ons afteleiden."

14 B.S. Albinus, *Dissertatio De Arteris*, 1: "Primum autem faciam in arteriis et venis intestiorum hominis, nactus opportunitatem artificio singulari, eoque eleganti, (...), expressas exhibendi."

15 Luyendijk-Elshout, "Morgagni, Ruysch, e Boerhaave," "The Elegant Anatomist."

as well: elegance contributes to the creation of knowledge, and thus has epistemic value. The epistemic value of elegance and perfection can be linked to the early modern equation of beauty and perfection to health and happiness. This equation derived from the work of the ancient Greek medic Galen (AD 129–c.210), which was still widely read among eighteenth-century anatomists such as Albinus.[16] Galen connected the sculptor Polycletus' idea that beauty is found in well-proportioned limbs to the stoic Chrysippus' statement that health is the symmetry of elements in the body.[17]

Following the growing importance of the experiment in the seventeenth century, throughout the course of the eighteenth century rational analysis of sensory perception and the understanding of the rules of beauty and perfection became increasingly important. This also meant that 'symmetry' or 'harmony' no longer sufficed as an answer to the question of what beauty and perfection entailed. The issues that arose from knowledge gained through sensory cognition as well as from the criticism of taste in the fine arts were first extensively addressed by the German philosopher Alexander Baumgarten in his 1735 *Meditationes philosophicae de nonnullis ad poema pertentibus* and in his 1750–58 *Æsthetica*. The latter title was derived from the Greek αἰσθητικ-ός, meaning material things perceptible by the senses (as opposed to immaterial, thinkable things), and which would quickly become a thoroughly ambiguous term. Few words have changed meaning so often, are so ambiguous and yet used so carelessly as 'aesthetics' and its derivatives, making it a potentially dangerous word with which to describe eighteenth-century practices and objects.

Disgust

The attempts to approach beauty and perfection rationally also implied that perfection might be understood through its negatives: the repulsive, the disgusting, and the ugly.[18] As mentioned earlier, there was no shortage of these negatives in eighteenth-century anatomy, as it could be a rather gory affair. Disgust was certainly something many people felt on first entering a dissecting room, as shows from a letter of 1801, written by freshman medical student

16 Nutton, "The fatal embrace," "The Fortunes of Galen."

17 Santing, "De Menselijke Canon," 486–487.

18 Just as ethics is commonly mistaken as a synonym for the understanding of 'what is right' instead of for the understanding of right *and* wrong, aesthetics is commonly mistaken for the understanding of beauty only: another reason to use the concept aesthesis, which entails both beauty (perfection, elegance) and ugliness (deformity, disgust).

Hampton Weekes to his father Richard Weekes. He recounted how he dragged an unsuspecting visitor along to see the dissection room:

> ... says he I never smell't such a stink in my life, began to spit about & hung back I could not help laughing, however I got him as far as the door just peep'd in & saw 3 or 4 subjects, there was only one young man there who was wishing to finish a subject, for bloodvessels,—he no sooner saw him but soon turn'd round keep'd on spiting & twisting about so we came down, tho: I did not tell you he said he was shure he should be sick, shure he should be sick.[19]

And although an anatomist might be expected to have been able to deal with the gore and mess of death and dissection more easily than others, the eighteenth-century Leiden anatomists were certainly not immune to the unnerving effects of our 'partibility.'[20] For example, the Leiden anatomy professor Wouter Van Doeveren (1730–1783) was frequently called upon for help with difficult births. In the case of a woman who had been in labour for three days, but whose child had died with the head stuck in the birth canal, Van Doeveren saw himself forced to sever and remove the head of the child with an instrument called a Levret's hook, named after its developer with whom Van Doeveren took classes in obstetrics while studying in Paris. Subsequently he could slowly pull out the remaining body of the child. Although the treatment was considered successful as the woman survived, Van Doeveren was appalled and exclaimed in a letter to a physician friend: "Nothing I desire more than to never having to use that again!"[21] But dealing with the disgusting insides of the body and with severed body parts was of course inevitable for anatomists and other medical men; paradoxically, it was of course also a manner in which to gain access to knowledge about the beauty and perfection of the human body. This, for example, shows in the writings of the Dutch lawyer and poet Willem Bilderdijk (1756–1831), who, though never having officially studied medicine, gave a detailed account of a dissection and used it to show that under all the bloody mess there is indeed something of profound beauty, and

19 Weekes quoted in Ford, "The Weekes Family Letters," 51.

20 Miller, *The Anatomy of Disgust*: "There are few things that are more unnerving and disgust evoking than our partibility. (...) Part of death's horror is that it too is a severance of body and soul and then, via putrefaction, of the body's integrity."

21 Van Doeveren. MS UvA: *Letter to J. Tak*, UB: HSS-mag.: Y 123.0.

tried to persuade his reader of both the necessity and pleasure of anatomical explorations.[22]

Why, then, do anatomical dissections and preparations, and sometimes even drawings, often evoke initial disgust, even though this often seems to subside after a while? First, it is useful to distinguish between 'core' and 'moral' disgust. The former is the kind of disgust that is viscerally responsive to foul or potentially contaminated things within close proximity; it is a protean emotion that may protect us from harm. The latter, however, has as its objects (or representations) people or behaviours that transgress social norms. These two forms of disgust are not always easy to separate.[23] Indeed, when first seeing an anatomical preparation, core disgust is evoked because, in general, dead bodies and separated body parts are not normally the kind of things we want to encounter in close proximity—even the suggestion makes us shiver. Simultaneously though, anatomical preparations may evoke moral disgust, as the social acceptance of putting human bodies and body parts in liquid-filled jars is limited to very specific purposes and circumstances. Only when we have satisfactory proof that this particular object does not represent the transgression of social norms, are we willing to accept it.[24]

Obviously, the level of satisfaction with the proof of social acceptability is, to a certain extent, something which is arbitrary, as shows from the fact that there are still wildly varying opinions on the acceptability of both the Ruysch preparations and modern anatomical exhibitions such as Gunther von Hagens' *Body Worlds*. Moreover, disgust can also be used as a means, for example, to entertain, moralize, or to educate, sometimes simultaneously. This was the case with a skull affected by syphilis kept in the Ruysch collection. It was the skull of a prostitute, and therefore a vivid moralizing example of the effects of promiscuous behaviour, slightly disgusting because of its origins and pathology, but entertaining to observe when held against the light as this clarified the thin patches in the bone, and served as a lesson on both human anatomy and venereal disease. After the skull broke—probably it had been dropped on one of the trips to a window or candle—Rusych turned it into an equally moralizing preparation by putting the fragments in a jar and positioning an infant's foot on top of them, an allegory of innocence trampling debauchery.[25] By transforming a broken preparation into a new preparation with a strong

22 Bilderdijk, "Antwoord Op De Vraag," appendix V, pp. 127–140.

23 Korsmeyer, *Fear and Disgust*, 4.

24 For more on deformity, ugliness and disgust in the eighteenth-century Leiden anatomical collections see Chapter 5.

25 Ruysch, *Thesauri Anatomica*, vol. VII (1707), 9, no. XIV.

allegorical meaning, Ruysch not only regulated the evocation of disgust in his visitors, he commodified the prostitute's skull for a second time. This brings us to the next aspect of aesthesis.

Commodification

Aesthesis in anatomy inevitably involved the related activities of commodification, domestication, and objectification. Commodification, turning something that is not by nature commercial into, or treating it as, a commodity, can be done to almost anything that occurs naturally: plants, animals, humans, but also water and landscapes.[26] Like slavery, prostitution, relics, headhunting, and corpse medicine, anatomical preparations are a form of commodification of the human body. Yet unlike most other forms of commodification of the human body, the creation of anatomical preparations was not aimed at the consumption of the body or its services, nor at stabilizing it for primarily symbolic, ritual, spiritual or decorative reasons (although these could also play a role). Making an anatomical preparation is first and foremost aimed at rendering knowledge, gained through sensory perception and experimentation with the body, stable and tradable. Sometimes this tradability was only of secondary importance, as for the many anatomists and collectors who made anatomical preparations primarily for their own research and teaching collections. In certain other instances it was the prime concern of the maker, as was the case with preparations that were made with the explicit intention of viewing by paying visitors and to be sold eventually.[27] This could also be the case when the maker wanted to ship preparations back to Europe. This shipping back of commodified bodies—human, animal, and botanical—ensured the maker stayed connected to the centre of knowledge.

In some cases, commodification also involved the explicit objectification of the human body, and was aimed at directing the gaze of the observer. The lace-rimmed sleeves and collars on the preparations of Ruysch and Albinus are an example of this: by covering up the stumps of limbs and necks, our eyes are diverted from the disgust-evoking sites of mutilation of the body,

---

26 On the commodification and exchange of naturalia in the early modern period see i.e. Cook, *Matters of Exchange*, Smith and Schmidt (eds), *Making Knowledge*, Dupré and Lüthy (eds), *Silent Messengers*.

27 For the advertisement and sale of anatomical preparations in the Low Countries in the late seventeenth and early eighteenth centuries, see Margócsy, "Advertising Cadavers," "A Museum of Wonders."

and are instead steered at the preparation itself.[28] Although clearly human, the preparation has become a commodity; an object that simultaneously is what it represents, although its meaning often stems from, and transcends, its materiality.[29] Moreover, the division of the body into parts ensures that the preparation does not primarily represent an individual. To use a term coined by Samual Alberti, the preparation has become a dividual: together with other preparations it forms a composite, multi-authored person, consisting of parts that originate from different places.[30]

On other occasions, commodification of the human body involved both the domestication of the exotic other, and of indigenous knowledge. We should keep in mind that commodification was almost never a simple accumulation, and direct distribution, of knowledge. Processes of transformation and manipulation were necessary to create stable, meaningful objects out of the ambiguous materials initially accumulated. This could also lead to rather incidental objects being redistributed and accepted as European 'universal' knowledge.[31] This seems to be the case with the late eighteenth-century preparations of 'African' and 'Asian' human foetuses in the Leiden collections: decorating them with beads and plants was simultaneously an inquiry into indigenous medical knowledge and a way to make a claim about their geographical origins.[32]

Thus it can be said that eighteenth-century aesthesis is a mode of thinking and working, a simultaneously philosophical and practical, mainly tacit, approach to questions of, and quests for, sensory perception-based knowledge, beauty and perfection, evoked by bodily material perceptible to the senses. Dealing with disgust takes up a special place within aesthesis, being both necessary because of the visceral character of practicing anatomy, and useful in directing users' perceptions of anatomical preparations and drawings. Commodification is inevitably an aspect of aesthesis as it is aimed at preserving and spreading knowledge about perishable bodies in lasting objects. Because of its tacit character, aesthesis is expressed primarily in the materiality of what remains of these anatomical practices: anatomical preparations, drawings, and every now and then in writings. Yet to understand why aesthesis is

---

28 See Chapter 4.

29 On this property of anatomical preparations see Rheinberger, "On Historicizing Epistemology," 233–234, and Huistra, "Preparations on the Move," 44–5. Brown, "Thing Theory," 4–5, describes the meaning of the object beyond its materiality as its 'thingness'.

30 Alberti, *Morbid Curiosities*, 71, 128.

31 Schmidt, "Accumulating the World," 133–4.

32 See Chapter 6 for more on these preparations.

characteristic of the eighteenth-century Leiden anatomical collections, clarification is needed on what distinguishes aesthesis from aesthetics.

Aesthesis, Not Aesthetics

As we have seen, aesthesis is an eighteenth-century word, used by medical men of the time to describe a certain sensibility to sensory impulses from the world around them. This is a valuable heuristic concept to understand the epistemic culture of which the eighteenth-century Leiden anatomical collections are the material remains. Yet what needs to be clarified is how aesthesis is different from the confusingly similar term aesthetics. The concept aesthetica is of course undeniably connected to Baumgarten, who published his *Aesthetica* between 1750 and 1758.[33] However, the emergence of the word 'aesthetica' in the mid-eighteenth century is not a bolt from the blue. In the preceding century, both the visual arts and natural philosophy had increasingly started to depend on perception and experiment, making the hand and the senses, especially the eye, important tools in gaining knowledge about the 'Book of Nature.'[34]

In order to be able to explain this combination of mental and practical activity in gaining knowledge, Baumgarten argued that there are two levels of cognition, namely logic—comparable to the Aristotelian scientia—and the lower level of aesthetics—comparable to Aristotelian artes—which is autonomous and has its own laws. The object of logic is to investigate the kind of perfection proper to thought, to analyse the faculty of knowledge; the object of aesthetics is to investigate the kind of perfection proper to perception.[35] Aesthetics is "the science of sensory cognition" (*scientia cognitionis sensitivae*).[36] From a Cartesian point of view, a science of perceptions is inherently paradoxical, as it is difficult, if not impossible, to submit perception to an exact and systematic treatment. However, as Eagleton remarks, there must be some way for reason to penetrate the world of perception while maintaining its absolute power—a delicate balance Baumgarten was attempting to establish within his work.[37]

It is thus in a sense partly based on the Baumgartian definition of aesthetics that aesthesis is used in this book—to designate an attitude that entails both

33 Baumgarten, *Aesthetica*.

34 See i.e. Smith, *The Body of the Artisan*.

35 Baumgarten, *Reflections on Poetry*, §§ 115–116.

36 Baumgarten, *Aesthetica*, § 1, 10–11, Beardsley, *Aesthetics*, 156–7.

37 Eagleton, *Ideology of the Aesthetic*, 15.

the investigation of the kind of perfection proper to perception and a science of, and based on, sensory cognition. But aesthesis is not the same thing as aesthetica or aesthetics. In order to clarify this, the historical and contemporary confusion linked to the term aesthetica has to be unravelled. The acceptance and use of the concept 'aesthetica' and its derivatives was neither quick nor widespread. In England as late as 1842 the word was still frowned upon by some as silly and pedantic.[38] Although Baumgarten coined the term, the rise of aesthetics as an independent discipline, a philosophy of art or more generally of beauty and taste, cannot easily be pinpointed to a specific place or period. Recently it has been presented both as particularly German and British, either in the nineteenth or the eighteenth century.[39]

The period 1711–1735 has been marked as "the foundational quarter century of the discipline," but by 1735 there certainly was not an academic philosophical discipline referred to as 'aesthetics' anywhere in western Europe.[40] A lot of eighteenth-century writing we would now describe as aesthetic theory was simply referred to by contemporaries as art theory or as discussion about taste, whereas theories referred to as aesthetics in the eighteenth century would now more likely be described as 'cognition theory.' Moreover, our contemporary discourse of aesthetics or philosophy of art has been determined largely by the work of Immanuel Kant, in particular by his third critique, the 1791 *Critique of Judgment*.

In 1787, Kant wrote in the first paragraph of the first part of his *Critique of Pure Reason*: "The science of all the principles of sensibility a priori, I call transcendental aesthetic," explaining in a footnote that, at the time, only the Germans used the word 'aesthetic' to indicate what others call the critique of taste. As this German discourse was widely translated in Dutch and read and discussed in the Dutch learned societies and salons in the early nineteenth century, it is likely that by that time educated Dutch also associated 'aesthetic' and its derivatives with questions of taste.[41] Kant goes on to argue that Baumgarten tried in vain to subject the criticism of the beautiful to the principles of reason, thus incorporating its rules into a science. He therefore felt it was advisable

38 De Bolla, *The Discourse of the Sublime*, Guyer, "The Difficulty of the Sublime," xi, Gwilt, "Aesthetic," 673.

39 Guyer, "The Difficulty of the Sublime," xi, claims that aesthestics as a discipline was predominantly shaped in eighteenth-century Britain, whereas Hammermeister, *The German Aesthetic Tradition*, x, defines it as a fairly recent development grounded in German philosophical thought.

40 Guyer, "The Difficulty of the Sublime," xi.

41 Onnasch, "Receptieonderzoek," 143–7.

> ... to give up the use of the term as designating the critique of taste, and to apply it solely to that doctrine, which is true science—the science of the laws of sensibility—and thus come nearer to the language and the sense of the ancients in their well-known division of the objects of cognition into 'aiotheta kai noeta' [perceived things and things known through reason].[42]

Subsequently however, Kant does not apply this suggestion entirely consistently himself. In his philosophy, he distinguishes between transcendental aesthetics on the one hand, which entails the aforementioned science of the laws of sensibility, and aesthetic judgment on the other, which entails judgments of taste and beauty.[43] Apparently, Kant eventually felt the judgment of taste to be so heavily dependent on the laws of sensibility that 'aesthetic' was needed as an adjective to designate these judgments correctly. It is not surprising that this subtle distinction between transcendental aesthetics and aesthetic judgment in Kant's overall complex metaphysics was lost on many. However, some of Kant's ideas on beauty, in particular the disinterestedness, both moral and otherwise, of judgments of taste as discussed in his 1791 *Critique of Judgement*, have been very deeply ingrained in our contemporary thinking about art and beauty. In fact, diffused neo-Kantianism is to be found everywhere in Europe for much of the nineteenth and early twentieth centuries.[44] Combined with the confused use of 'aesthetics' both in the past and the present, it can therefore be hard to understand the pre-Kantian meaning of aesthetics in its own right. The main subject of this book focuses on the work of people who nonetheless fall mainly into this pre-Kantian period. In addition to this, it is worth noting that the popular and philosophical reception of Kant's work in the Netherlands did not gain momentum until well after 1790.[45]

In the course of the nineteenth and twentieth centuries the term 'aesthetics' has come to be used as a synonym for philosophy of the arts, although 'aesthetic' and its derivatives have recently even been adopted as "a sentimental archaism for the ideal of beauty," removing the word still further from its

42 Kant, *Kritik Der Reinen Vernunft*, 62, footnote page 920, English quote from the Meiklejohn translation.

43 On transcendental aesthetics see Kant's *Critique of pure reason*, on aesthetic judgment see the *Critique of Judgement*.

44 Cheetham, *Kant, Art and Art History*, 4.

45 Onnasch, "Receptieonderzoek," 143–7.

original etymological and philosophical meaning.[46] The meaning of the word in the eighteenth century was possibly even more ambiguous than it is now, and this 'new discipline' certainly did not have clear boundaries at the time. Surprisingly, even historians of art and science recently have been imprecise in applying the term aesthetics to historical objects, methods and theories. When discussing objects or works of art, 'aesthetics' is used to refer to visual features, particularly stylistic or decorative elements. The art historian Julie Hansen, for example, speaks of the 'aesthetics' of the work of the Amsterdam anatomist and man-midwife Frederik Ruysch, taking the paintings of the surgeons' guild as a starting point, and writes: "... Ruysch should be appreciated first and foremost as an artist."[47] This is rather anachronistic, as she uses the words 'art', 'science' and 'artist' in their twentieth century meanings, without explaining that they meant something quite different at the time, as did 'aesthetic'. Hansen ascribes post-Kantian ideas of the artist (autonomous, creative, difficult personality) to Ruysch and the painters of his time, ignoring the fact that 'the artist' was an entirely different concept in the late seventeenth and early eighteenth centuries, and that art and science were largely undivided territory in early modern times.[48]

The separation between art and anatomy, artist and anatomist in the eighteenth century was not as straightforward as our contemporary categories suggest. As Smith has pointed out, 'arts' has long referred to skilled manual operations in general, which we would now categorize mainly as 'crafts' and not quite as the fine arts we now think about upon considering the word.[49] Therefore, 'artists,' or artisans, as Smith prefers to call them, who made anatomical illustrations, were often not just visual artists, but also jointly poets, tradesmen, administrators, or anatomists. Their commissioners often considered them skilled craftsmen, not the autonomous, unpredictable genius we often take 'the artist' to be. This is simply because the problematic concept of the artistic 'genius,' in the sense of the visionary, powerfully possessed artist, did not emerge until the late eighteenth century: it is a thoroughly Romantic concept. Indeed, for most of the eighteenth century, it was 'ingenuity,' natural

46 Baumgarten, *Aesthetics, Reflections on Poetry*, Beardsley, *Aesthetics*, 156–9, OED online, consulted 17 August 2009.

47 Hansen, "Resurrecting Death," 678.

48 Also see Jorink and Ramakers, "Undivided territory."

49 Smith, *The Body of the Artisan*, 27.

cleverness and inventiveness necessary for the craftman's skills of both natural philosophers and artisans that were appreciated most.[50]

Similarly, anatomists were never just anatomists; they mostly thought of themselves primarily as natural philosophers and their work included activities we would associate with physicians, astronomers, physicists, philologists, theologians, philosophers, governors, publishers, collectors, and agriculturalists. For example, Leiden anatomy professor Wouter van Doeveren was also an avid private collector, a city physician, a forensic medic, and one of the central figures in the Dutch smallpox and cattle-plague debates. His colleague Sebald Justinus Brugmans held doctorates in philosophy and medicine, taught logic, metaphysics, natural history, astronomy, botany and medicine, was head of the national Military Medical Services twice and strongly influenced national policy through his reports on cattle-plague, his proposal for the foundation of a veterinary college, and his contributions to the first national pharmacopeia. Moreover, he was one of the founders of the new Leiden drawing academy and owned a large private art collection. In this light, our modern categories of art and medicine become fluid, maybe even superfluous.

Despite this recent rethinking of categories, historians of science are also still remarkably careless with the use of the word aesthetic in their work. Even when methods or theories are referred to as 'aesthetic' the implicit meaning of the word often seems to be solely 'beautiful' or even 'artistic,' as appears to be the case when Jardine, for example, discerns three types of aesthetic appraisal that may be counted in favour of a theory or hypothesis, such as those which 'bring out' certain aesthetic virtues in the (representations of) phenomena that they try to explain. Not once is it made explicit just what exactly is meant by 'aesthetic virtues,' although the attentive reader suspects it has something to do with pleasant visual elements.[51] Ball too, in his recent work on the beauty of chemistry, explains fairly well what he means by 'beauty,' but subsequently uses 'aesthetic' as a synonym for beauty without further consideration.[52]

This confusion explains why we should not use 'aesthetica' or 'aesthetic(s)' when discussing the eighteenth-century Leiden anatomical collections: the first is too exclusively Baumgartian, the second too eroded and thus empty and confusing. Nevertheless, the fact remains that the Greek αίσθητιϰ-ός, material things perceptible by the senses, the dual meaning of investigation of the kind

50 Schaffer, "Genius in Romantic Natural Philosophy," 82, and Shapin and Schaffer, *Leviathan and the Air-Pump*, 129–131. In Latin, 'genius' originally applied to the personal deity that watched over an individual's fate.

51 Jardine, *The Scenes of Inquiry*, 195–206.

52 Ball, *Elegant Solutions*, 8, 10, Chapter 10.

of perfection proper to perception, and a science of sensory cognition found in the Baumgartian 'aesthetica,' played a central role in the genesis of the eighteenth-century Leiden anatomical collections. Even though aesthetica and aesthetics do not seem to appear in the vocabulary of the eighteenth-century anatomists, I argue that this is one of those cases in which something is so obvious it does not need to be mentioned explicitly by those involved.

Anatomy was habitually referred to as an art, in the Aristotelian sense by its practitioners well into the nineteenth century, a skilled manual and sensory practice paired with the more mental and rational *scientia* of physiology to map and understand the human body.[53] Anatomical knowledge and skills were also of great importance for visual artisans such as painters and sculptors, and the skills and knowledge of those artisans were in turn indispensable for anatomists in mapping the body and developing knowledge. It was not unusual, therefore, for the anatomist and artisan to frequently be one and the same person. The most vivid Dutch example was again Petrus Camper, who not only published on human and animal anatomy, but also on drawing techniques, fossils, shoes, and megalithic graves, and skilfully illustrated almost all his work himself. Even in the cases where the anatomist could not compete with the draftsman, he was almost always passionate about depicting anatomy, possessed at least basic drawing skills and was often closely involved in both drawing academies and learned societies, institutions which were frequently found under the same roof by the late eighteenth century.[54]

Upon closer investigation aesthesis not only transpires in the anatomical preparations, but in all facets of the anatomists' work and lives, and explains the fascinating, though confusing, diversity in materials and visual appearances these collections display. The eighteenth-century Leiden anatomical preparations are the result of a profound eighteenth-century aesthesis which includes both the continuous use of sensory perception as a source of knowledge and a permanent tacit quest for finding and understanding beauty, as well as dealing with the disgusting aspects of anatomy, along with the desire to commodify and objectify the human body. Notwithstanding the often confusing and confused meanings of its paronyms, the term will become precisely requisite for what is needed in understanding the eighteenth-century Leiden anatomical collections, and may prove to be equally useful in understanding similar collections from the same period.

---

53 See Cunningham, "The Pen and the Sword," on this divide and how it shifted.

54 A good example is the strong involvement of the Amsterdam professor of anatomy Andreas Bonn (1738–1817) in the establishment and programming of the Amsterdam society for arts and sciences, Felix Merites (see Bonn, *Redevoeringen*, 1790).

Aesthesis as Analytical Category for an Epistemic Culture

What makes aesthesis such a central concept in understanding the Leiden anatomical discourse of the long eighteenth century? Obviously, eighteenth-century natural philosophy and anatomy were subject to changes. Answers to questions concerning the fundamental entities the universe is composed of, and how these interact with each other and the human senses, were manifold. Sensory perception, particularly vision, was not securely established in relation to notions of truth-to-nature and fact.[55] Nor was the academic community any longer unanimous in acknowledging particular past achievements in the field of anatomy as the foundation for its practice—the eighteenth century would become the century in which the ancients would lose their position as the first and foremost authority of reference for anatomists. Whereas Bernard Siegfried Albinus continues to preface virtually all his written work with perfunctory references to classical canon in anatomy, starting with Galen, his successors had virtually abandoned this custom by the end of the eighteenth century.[56] This led to many controversies, some of which lingered for decades.

Moreover, Dutch natural philosophy and anatomy were strongly rooted in traditions of (self-) representation, craft, and trade.[57] Thus, conversion of impressions and sensations into imagination and ideas through active awareness, or consciousness of, the role of beauty, perfection, and taste therein was of great importance to natural philosophers in Europe in the middle of the eighteenth century. This is apparent not only in the rationalist oriented work of Baumgarten, with its strong focus on the cognitive aspects of sensory perception, but also from the more emotively reasoned work of Mendelssohn, who investigated the pleasure derived from sensory-perceived objects.[58] Sense perceptions were the most important source for gaining knowledge. Regardless of whether most attention was paid to the cognitive or the pleasurable aspects, both are found in virtually all philosophical work of the period, such as Hume's 1748 *An Enquiry Concerning Human Understanding* and Offray de la Mettrie's *l'Homme-Machine* from the same year, as well as in the work of the Dutch phi-

55 On scientific images and the historical development of concepts like truth-to-nature, trained judgement and objectivity, see Daston and Galison, *Objectivity*. On the relationship of sensory perceptions and fact in the seventeenth century, also see Clark, *Vanities of the Eye*.

56 Cunningham, *The Anatomical Renaissance*.

57 Alpers, *The Art of Describing*, Pickstone, *Ways of Knowing*, 2001, 64–7.

58 Hammermeister, *The German Aesthetic Tradition*, 13–14. Collingwood, *The Principles of Art*, 306, Graham, *Philosophy of the Arts*, 34.

losopher Frans Hemsterhuis, a close friend of Leiden-trained anatomist Petrus Camper. Hume thus reflected on anatomy:

> The anatomist presents to the eye the most hideous and disagreeable objects; but his science is useful to the painter in delineating even a Venus or an Helen. While the latter employs all the richest colours of his art, and gives his figures the most graceful and engaging airs; he must still carry his attention to the inward structure of the human body, the position of the muscles, the fabric of the bones, and the use and figure of every part or organ. Accuracy is, in every case, advantageous to beauty . . .[59]

These sensory perceptions were hugely important for Hume as he held that ideas are nothing more than the products of "compounding, transposing, augmenting or diminishing the materials afforded us by the senses and experience."[60] La Mettrie mentions beauty and the beautiful twelve times in total in a text of merely 20,000 words, and in admiration of the physician Johann Conrad Amman, who taught deaf-mutes to speak, he writes:

> He who has discovered the art of adorning the most beautiful of kingdoms [of nature], and of giving it perfections that it did not have, should be ranked above an idle creator of frivolous systems, or a painstaking author of sterile discoveries.[61]

Near the end of his treatise he concludes: "Experience has thus spoken to me in behalf of reason; and in this way I have combined the two."[62] Hemsterhuis, although lacking a systematic philosophy, also reasons along the same lines. In his 1769 letter on sculpture, he had already observed that beauty is found in those objects of perception which provide the soul with the greatest number of ideas in the shortest time possible, and in his 1771 work on man and his rapports goes on to argue that:

> The science or knowledge of man consists of ideas which he acquires through the senses and from those of rapports which exist between those ideas. The first are isolated, and represent isolated objects, the others

59 Hume, *An Enquiry Concerning Human Understanding*, 4 (section I, §5).

60 Ibid., 11 (section II, §13).

61 La Mettrie, *L'homme Machine*, 102.

62 Ibid., 149.

> derive from the existence of a certain number of the first, which the intuitive faculty can grasp at once.[63]

Five years later, in his treatise on the reality of appearances, Hemsterhuis boldly concludes: "I sense and therefore I am."[64] Once again, it appears that sensory perception is inextricably connected to both questions of beauty, or perfection, and to gaining knowledge in the eighteenth-century mind. This dual philosophical stance had far-reaching practical implications, as we will see in the work of the Leiden anatomists.

It is likely that the aforementioned philosophical works were fairly well known among scholars and educated men, disseminating on the social level through the particularly eighteenth-century phenomenon of the 'society.' Learned, philosophical, and royal societies for the advancement of the arts and sciences sprang up all over Europe, often as reactions against the perceived fossilization of the universities. In the heterodox Netherlands with its well-established print industry and book trade and lack of intellectual nationalism, the universities functioned fairly well in comparison to those in the neighboring countries. The room for experimental research and Cartesianism in Dutch universities, most notably Leiden University, might explain the fact that Dutch non-university academies were few and not as grand and modern as those in Paris and London.[65] The first and only notable learned society in the eighteenth-century Netherlands was the Hollandse Maatschappij der Wetenschappen (Dutch Science Society) of Haarlem. This society was only established in 1752 and acknowledged as 'societas litteraria' by the States of Holland under protest from Leiden University in 1760, as the University feared the loss of its unique status would damage its reputation.

However, educated men, (and occasionally women too), soon started founding societies and associations all over the country, independent of the universities. These were mostly utilitarian in character, that is, they aimed at informing and educating citizens and their ambit was mostly local. In Amsterdam alone,

---

63 Hemsterhuis, *Wijsgerige Werken*, 91, 501–503.

64 Ibid., 135. It has been suggested to me that Hemsterhuis took this idea from the work of Johann Gottfried von Herder (1744–1803). However, as far as I have been able to trace it, Von Herder first ventilated a similar statement in his 1778 *Plastik* (1809 ed., 244), so three years later than Hemsterhuis (also see Pochat, *Geschichte der Ästhetik*, 429, Barasch, *Modern Theories of Art*, 165–171). In fact attention to sensory perception as a source of knowledge is fairly common in late eighteenth-century philosophy, so even though Hemsterhuis' thought was probably not original, it was an accurate reflection of the epistemological philosophy of his age.

65 McClellan, *Science Reorganized*, xxii.

over ninety reading circles, freemasonry lodges, societies for music, medicine, mathematics, and archery were counted in the second half of the eighteenth century.[66] Some of the learned societies, mainly outside the Netherlands, were large, formal structures, such as the English Royal Society, but everywhere there were numerous small, informal, local groups meeting on a regular basis to discuss art, sciences, politics, and sometimes even religion. These groups were essential for the enlargement of intellectual enterprise and its integration in society.[67]

Most anatomists, both in the Netherlands and abroad, were members and attendants of such societies and salons and maintained numerous international contacts with colleagues and acquaintances. Foreign visitors often not only visited a university, but also attended society meetings, and academics and men of learning visited each other at their homes to discuss the latest developments in their field. Even students used letters of introduction written by supervisors to pay home visits to academics in foreign cities. Wouter Van Doeveren (1730–1783), professor of anatomy at Leiden University from 1770 until 1783, spent about a year in Paris after his studies in Leiden in 1752–1753. There he not only took classes with famous anatomists, he also visited degree ceremonies, theatres, and famous scholars in their homes, such as Nollet and Reaumur.[68] It is often quite difficult to trace these contacts as many of them were either never registered or any documentation on them was lost, but it is clear that this entanglement of academic, societal, and social exchanges made for a vibrant culture of knowing.

Although some have dismissed the eighteenth century as a period of decline with little international significance for Dutch academia, Leiden University was still a fairly respected institution in the second half of the eighteenth century.[69] The Leiden anatomists were part of extensive international networks maintained through letters, visits, and the exchange of gifts, books, and favours, as will become clear throughout this book. They owned and discussed contemporary literature from the Netherlands and abroad covering a wide variety of topics. For example, in Albinus' library we find works by Leibniz, Hunter, Buffon, Boyle, Descartes, and Van Doeveren owned copies of works by Baumgarten, Mendelssohn, Hume, Offray de la Mettrie and Buffon.[70]

66 Bots, *Universitaire Gemeenschap Tot Academische Kring*, 12–15, McClellan, *Science Reorganized*, 123–5.

67 Ibid., xxii.

68 Van der Zwaag, "Wouter Van Doeveren," 15.

69 I.e. Jacob and Mijnhardt (eds), *The Dutch Republic*, Israel, *The Dutch Republic*, 1049–1051.

70 Van Doeveren, *Bibliotheca Doeveriana*, 226–231.

Apart from these exchanges, the relationship between observation and experiment shifted several times from the early seventeenth until the mid-nineteenth century.[71] Similar tendencies are noticeable in eighteenth-century anatomy, both in Leiden and elsewhere. For example, by the end of the eighteenth century, to simply claim that a skull was from a particular country or area was no longer enough for some anatomists to prove the existence of this region and its inhabitants, and they resorted to taking and filing endless measurements of skulls to distinguish certain categories into which particular 'species' of humans could be made to fit. Another example of an increasing suspicion of singular and unverifiable sensory perceptions among anatomists in the second half of the eighteenth century can be found in the increase in collections containing multiple specimens with a similar pathology or birth defect.[72] However, the scope of this work is limited to the eighteenth-century Leiden anatomical collections, and to aesthesis, which offers a unique opportunity to access and understand the particular materiality of the preparations.

What difference does aesthesis make in writing a history of science, or a history of art or medicine for that matter? Why is it a unique, new perspective, and what does it add to this particular topic, the history of the eighteenth-century Leiden anatomical collections? As noted earlier, aesthesis in the present book is used in its eighteenth century, non-disciplinary sense, to refer to an epistemic culture, a mode of thought and practice, a tacit approach to questions of, and quests for, sensory perception-based knowledge, beauty, and perfection evoked by bodily material perceptible by the senses. This key concept, to my knowledge, has not before been clearly defined, let alone used in writing a history of science.

That the notion of aesthesis is essential to, and an enhancement of, the history of science and medicine is demonstrated by the fact that it benefits both the general way in which we deal with, and understand, historical medical and scientific objects, as well as historical notions of beauty and perfection, and more particularly by its being a key with which to unlock anatomical preparations, together with its practices and objects, in the eighteenth century. It will appear from the cases discussed in this book that the seeking and displaying of beauty and perfection, as well as sensory perception as being a source of knowledge, play essential roles in Leiden anatomy in this period. As this seeking and displaying are almost entirely tacit and material, the concept of aesthesis is essential in identifying and understanding them.

71 Daston, "The Empire of Observation," 82.

72 See Chapters 5 and 7 of this book for more on these examples.

Second, although attention to the role of sense perception in the history of science has increased in recent years, the unique eighteenth-century combination of this with the search for beauty and its rules can only be explained through aesthesis. For example, Riskin has successfully proposed the notion of 'sensibility' as key to understanding French empiricism in the same period, but this concept does not cover the quest for beauty and perfection so obvious in Leiden anatomy, and thus lacks the breadth required for this study.[73] Schickore has persuasively shown that the visual culture of the microscope, contrary to that generally maintained in existing accounts of eighteenth-century microscopy, indeed played a considerable role in anatomical research and in the validation of concrete results.[74] However, Schickore's account does not explain the diversity of visual features of collections such as those in Leiden, nor does it take into account that aesthesis is as much about the other senses as about vision, and that beauty and perfection are as equally involved as matters concerning sensory perception. Sensibility and visual culture might be useful notions for understanding French empiricism and eighteenth-century microscopy, but it is impossible to understand the Leiden collections without the hybrid notion of aesthesis, which covers both the importance of the entire spectrum of sensory perception in understanding human anatomy and the issues of beauty, knowing, and taste arising in and from this discourse.

Finally, this coining of the concept of aesthesis and the subsequent inquiry into the role of aesthesis in eighteenth-century Leiden anatomy is based on, and will add to, recent historiographic exercises on notions such as agency, object, and locality. The centrality of the unheimliches historical object and the study of evolving and mutating epistemological concepts were inspired by Daston's *Things that talk* (2008) and her earlier work on facts and objectivity.[75] This is clear from my examination of the confused idea of 'the aesthetic,' which we have long used to organize ideas of beauty and perfection in the history and philosophy of science. By coining the concept of aesthesis, I take the exercise a step further. My proposition of aesthesis as an analytical and descriptive concept for an epistemic culture and objects stems from the inquiry I made into the use of the term 'aesthetics.' This theoretical exercise is inspired by, and is a contribution to, both material culture studies and the historical epistemology Hacking advocates: the need to examine the ideas we use to organize knowledge and inquiry, and to propose, advocate, or refute theories

73 Riskin, *Science in the Age of Sensibility*.

74 Schickore, *The Microscope and the Eye*, 6–82.

75 Daston, "The Image of Objectivity," Daston and Park, *Wonders and the Order of Nature*, Daston (ed), *Things that Talk*.

of knowledge.[76] Last but not least, I hope that this book will reveal the preparations in the eighteenth-century Leiden anatomical collections in all their glorious complexity. After all, this book is a historical study of objects of science and medicine as produced by people with bodies, situated in time, space, culture, and society, and struggling for credibility and authority, to paraphrase the subtitle of Shapin's recent monograph.[77]

In this chapter, aesthesis has been defined as a useful historiographical analytical category to describe an eighteenth-century epistemic culture. Aesthesis was largely tacit, based in sensory perceptions and materiality, and its most important characteristic is that its epistemic culture was strongly concerned with seeking and bringing out beauty, perfection, and elegance in both its objects and its makers. In addition, it also entailed the quest for beauty and elegance for dealing with the inevitable disgust connected with doing anatomy, and the commodification, objectification, and domestification of the other, both exotic and familiar. This chapter has explained that although related, aesthesis is fundamentally different from the confused and vague term 'aesthetics.' Although aesthesis might not be particular to eighteenth-century Leiden anatomy, it certainly strongly influenced the material characteristics of the Leiden anatomical preparations, and it is very likely that Leiden acted as an important hub in the spreading of it. Finally, the concept of aesthesis emerged from, and contributes to, the field of historical epistemology, and is essential in understanding the eighteenth-century Leiden anatomical preparations as material objects that stem from a particular epistemic culture, and which is communicated in their materiality.

---

76 Hacking. *Historical Ontology*, 8–9.

77 Shapin, *Never Pure*.

CHAPTER 3

Quicksilver Anatomy: Exploring the Lymphatic System with Mercury

Quicksilver was an appreciated but notoriously challenging material for anatomists in the long eighteenth century. Preparations injected with a mixture of mercury and hardening agents gain an attractive silvery shimmer if the injection is successful. This chapter analyses the use of quicksilver in anatomical preparations in Leiden in the eighteenth century, revealing aesthesis in all its complexity. Quicksilver-injected preparations literally are eyecatchers. Housed in the anatomical museum of the Leiden University Medical Center, this preparation is briefly described in the most recent catalogues as 'an exceptionally fine specimen,' and attracts the visitors' attention straight away with its silvery shimmer.[1] The curator of the collection in the 1950's, Ms Elshout (the later professor Luyendijk-Elshout) attributed it to Eduard Sandifort (1742–1814), professor of anatomy at Leiden University from 1771 to 1814. There are other mercury-injected preparations of the lymphatic system in the collections too, some wet, some dry, some of separated lymphatic vessels, others of lymphatic vessels in situ, of the heart, the liver, the penis, and the intestines. This chapter explores whether all these preparations were made by Sandifort, why these body parts were chosen and prepared with mercury, how the preparations were used, and what they meant to their contemporaries, while paying specific attention to the material aspects of aesthesis.

Moreover, I argue that the initial choice for mercury as injection mass for the lymphatic system was no coincidence, but the result of both the materiality and the meanings of mercury in the second half of the seventeenth century. Quicksilver to the late seventeenth-century anatomist was easy to discern, it was the penetrating, cool, wet opponent of dry, hot sulphur, it was cleansing, resurrecting, influenced the hands and was a source of eloquence. This chapter will explain those meanings and how they changed, as the subsequent reappearance of mercury as an injection mass for anatomical preparations in the second half of the eighteenth century was possible because of subtle changes in the meanings of mercury. The materiality of mercury in these preparations

1 Elshout, *Het Leidse kabinet*, LUMC catalogue: "Buitengewoon fraai praeparaat van de lympheklieren uit de regio inguinalis. De klieren zijn gepraepareerd op de fascia abdominus superficialis, die geplet en gehard is. De lymphevaten zijn met kwikzilver opgespoten, waardoor een zeer fraai effect verkregen is."

 | DOI 10.1163/9789004262775_004

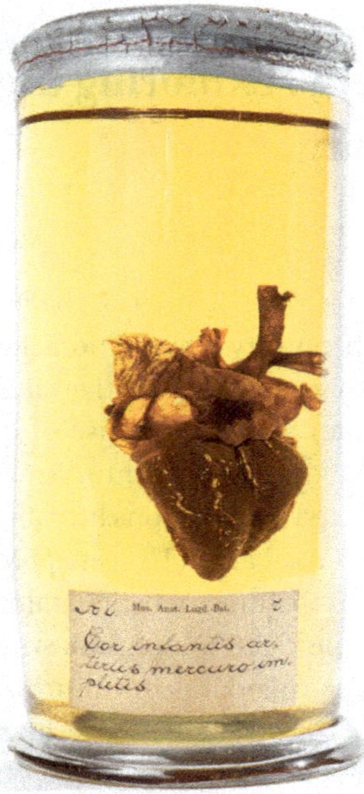

ILLUSTRATION 3.0 *LUMC Al0007. Heart injected with red wax and mercury. Probably from the collection of Eduard Sandifort (1742–1814). © Anatomisch Museum LUMC 2012. Photographer: Arno Massee.*

gives us access to, and is better understood through, a complex world of anatomical and medical knowledge and practices. These objects are, after all, evidence of complex social relationships; they are simultaneously products and modellers of a distinct epistemic culture.[2] As Klein and Spary have recently pointed out, materials can be challenging things that "provoke their investigators to expand and refine their activities and understanding." Materials speak irresistibly, and not only through their interpretation and representation.[3] In the case of mercury, we will see its agency and meaning are in its colour, its effect on the human body, and its volatile and intangible character. The mercury-injected preparations made by Sandifort were the products of typical eighteenth-century aesthesis.

2 Harvey, "Introduction: Practical Matters," 5, 12.

3 Klein and Spary, "Introduction: Why Materials?," 9.

The Mysterious Origins of the Leiden Mercury Preparations

Why and how were these preparations originally made and perceived? That question is in fact very hard to answer. A.M. Elshout attributed three preparations to Eduard Sandifort in the 1950's stating that some of these preparations had originally been attributed to the seventeenth century Leiden professor of anatomy, Nuck (1650–1692), but that it was more likely they were Sandifort's work as Nuck only left dry preparations, and because they were made exactly according to Sandifort's technique.[4] This statement is interesting enough, but unfortunately Elshout in her catalogue records did not refer to sources that attributed the preparations to Nuck, nor did she elaborate on what 'Sandifort's technique' purports to be.[5] Curiously, in her 1952 PhD thesis on the eighteenth-century Leiden anatomical collections, Elshout concludes that Nuck's preparations were most likely all removed from the collections in 1721, and that Sandifort only described two dry preparations by Nuck in his 1793 listing of the Albinus collection.[6] Moreover, she excluded the 38 preparations that she thought to be Sandifort's work from further discussion in her thesis as these were, strictly speaking, not a part of the eighteenth-century Leiden University anatomical cabinet: they were not listed in the four volume *Museum Anatomicum* catalogue made of the Leiden anatomical collections by Eduard Sandifort and his son Gerard in the years between 1793 and 1835, nor are they described or attributed to Eduard Sandifort in any other catalogue.[7] After briefly describing the mercury-injected preparations and some others, Elshout concluded that "the investigator gets the impression that these preparations have been made more to flatter the eye than for a scientific purpose, although the style of preparing is entirely different than that of Ruysch"—a style she characterizes elsewhere as 'excessive.'[8]

As these preparations have been sealed with the same characteristic grey lacquer as other preparations from Sandifort's collection, it is fairly safe to assume that they were indeed Sandifort's work. But would Sandifort really

---

4 These preparations now are LUMC catalogue numbers Ag0020, Ag0021 and Ag0022.

5 Catalogue records Elshout, *Het Leidse kabinet*, LUMC catalogue.

6 Elshout, *Het Leidse kabinet*, 14, Sandifort, *Museum Anatomicum*, vol. 1, 89, describes no. 407–408 as follows: "CCCCVII. Portio intestine cum portione mesenterii, vasis nonnullis mercurio impletis. Nuckii ut videtur. CCCCVIII. Portio intestine, cujus vasa lactea impleta mercurio. A Nuckio, aut Swammerdammio, ut videtur." This suggests he was not entirely certain about the origins of these preparations either.

7 Elshout, *Het Leidse kabinet*, 23–4, Sandifort and G. Sandifort, *Museum Anatomicum*, vol. 1–4.

8 Ibid., 24, 57.

have gone to the trouble of making such refined preparations only to 'flatter the eye'? Why would he have made preparations imitating a style and technique apparently last used by a predecessor almost a century earlier, and why would those preparations subsequently have been virtually ignored by his contemporaries and successors? In order to come closer to answering any of these questions, we have to delve into the history of making mercury- injected anatomical preparations, in particular the practice of so doing by Antony Nuck. Nuck is generally believed to have been the first anatomist who successfully used mercury to inject the lymphatic system, and to create lasting mercury-injected preparations.[9] So why did Nuck create these preparations, and why was he so successful at it?

Injecting and Preparing as Experiment

Antony Nuck (*1650) graduated from Leiden University in 1677 with a dissertation on diabetes, and subsequently practiced medicine and anatomy for a decade in The Hague. In 1687 he was appointed professor of anatomy at Leiden University, a post he would hold until his early death in 1692. In his time, he was famous for his numerous dried preparations of lymph glands, arteries, veins, eye and saliva ducts, many of which were injected with a mixture of mercury, lead, and tin (the lymph glands), and coloured wax (arteries and veins). In his lectures, Boerhaave mentioned how impressed he was as a student by the sight of mercury-injected lymph vessels mounted on wooden boards at Nuck's house.[10] Preparations of mercury-injected lymphatic glands were an admired novelty in the late seventeenth century for two reasons: the development of lasting anatomical preparations from injecting experiments was in full swing, and the lymphatic glands were a largely mysterious bodily structure.

Anatomy was primarily experimental in seventeenth- and eighteenth-century Leiden. In the 1660s, anatomical structures and functions such as the lymphatic system were discovered primarily through injections with air and coloured fluids. The instrument of choice for these injections was the syringe. Reinier de Graaf (1641–1673), a Delft physician who had studied medicine in Leiden, improved the syringe, and even wrote a short treatise on its use in anatomy: *Korte beschryving van 't gebruyk der spuyt* (1668). In it, he presented his

9 On Malpighi: see Cole, "The History of Anatomical Injections," 293, quoting Malpighi. The most convincing study to point out Nuck as the first who used solidifying mercury to make lasting preparations of the lymphatics is Helm and Stukenbrock, *Anatomie*, 72.

10 Boerhaave, Praelectiones Academicae, 576.

ILLUSTRATION 3.1 *LUMC Ak0006. Mercury-injected lymph vessels. Probably from the collection of Antony Nuck (1650–1692). This preparation on a slate of glass was originally mounted on a dark wooden board. In the 1990s it was removed from the board and put in a jar because of health and safety regulations. © Anatomisch Museum LUMC 2012. Photographer: Arno Massee.*

own version of the syringe: a copper tube with a piston, onto the front of which tubes of different shapes could be screwed. With this instrument, the anatomist could inject liquids of different colours into the veins and arteries of dead bodies to demonstrate to students which arteries supplied which intestines, and how blood flowed through the arterial system. Although most subjects used for these demonstrations were death, the syringe was also very useful in the vivisection of animals, as it provided the anatomist with an opportunity to show the audience bodily functions in vivo. For example, De Graaf demonstrated the flow of the blood and chyle in bound and gagged living animals by replacing it with injections of water.[11] De Graaf also saw great opportunities for

11 De Graaf, *Korte Beschryving*, 666–7.

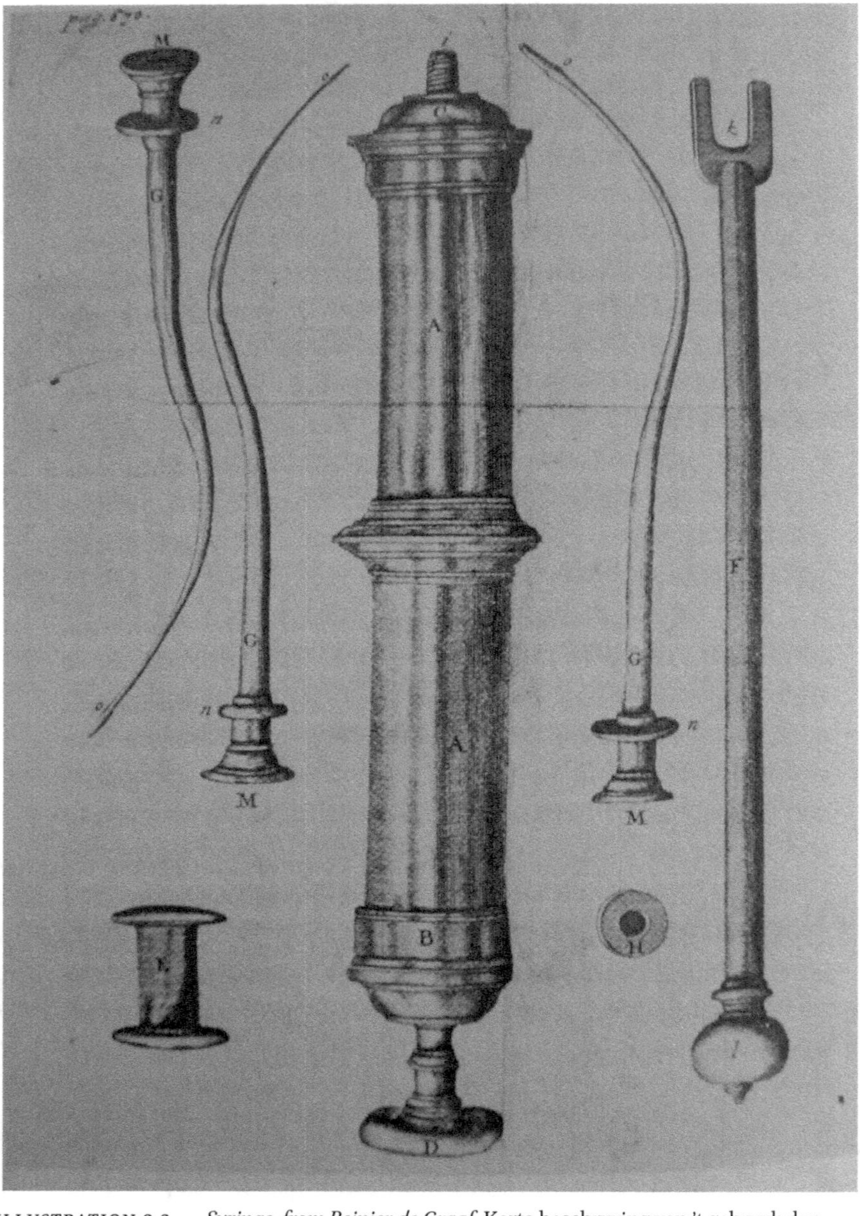

ILLUSTRATION 3.2 *Syringe, from Reinier de Graaf,* Korte beschryving van 't gebruyk der spuyt *(1668). Reproduced from the 1989 facsimile. © Reinier de Graaf Stichting, Delft, 1989.*

anatomists to discover thus far 'hidden things,' which would be of benefit to everyone and would make the anatomist himself 'immortal.'[12]

De Graaf apparently selflessly shared his discovery with his contemporaries, enabling them to make themselves immortal too. To him, this made perfect sense: patents and copyright did not exist, and the usual manner of purchasing instruments for anatomists was to take an illustrated description such as De Graaf's to a blacksmith and have the instrument made there. By publishing his new find, De Graaf knew that he would at least get the credit for this particular model of syringe, although he was unlikely to gain anything from it financially. More importantly, only having the description of such an instrument and its use did not in itself enable an anatomist to perform injection experiments successfully. Animal and human bodies and body parts are resistant materials to work with; it is hard to discern small veins to begin with, let alone injecting them successfully with coloured liquids. Too little pressure, and nothing happens, too much pressure, and the fluid destroys structures instead of highlighting them. Use the wrong colour and hardly anything will be seen of an otherwise successful injection.

Like any artes in the Aristotelian sense—anything that requires skilled handiwork- early modern anatomy required endless practice.[13] De Graaf appears to have taken this for granted, but in other anatomical handbooks we do find warnings, such as in Blankaart's 1678 *Nieuw-hervormde Anatomie* (Newly Reformed Anatomy). He wrote that before anything could be accomplished in the art of healing, knowledge of human anatomy should be acquired, "not from books... but from practice, to acquire a skilful hand."[14] Over a century later, in the introduction to the 1790 *Anatomical Instructor*, Pole still notes that to "Those who wish to become complete Anatomists, I must urge to devote a sufficient time at the usual places of instruction; to them, I hope, this treatise will be found, what I mean it to be, an useful companion and assistant."[15] Later on, he states that 'imitating nature by colours' [in preparations] is an art in itself, distinct from, but very necessary for anatomy. Once again, this can only be learned through practice.[16] This shows that descriptions of practical finds such as De Graaf's improved syringe could only partly spread new anatomical experimental knowledge. In order to truly learn the techniques, an anatomist

12 Ibidem, 14.

13 Also see Smith, *The Body of the Artisan.*

14 Blankaart, *Anatomia Reformata*, 1688, 2: "... en dat niet uit boukken (....), maar uit eigen ouffeninge, om van tyd tot tyd een vaardige hand-greep the krygen."

15 Pole, *The Anatomical Instructor*, xiv.

16 Ibid., 250.

needed to experiment with them, and only the most persistent and most skilled practitioners would eventually attain satisfying results.

Although the experiments and discoveries of De Graaf and his contemporaries with injections of coloured liquids were very innovative in terms of making structures and functions visible, they did not provide lasting preparations, as the injected substances did not solidify and preserve the injected structures. Creating lasting injected preparations was something many anatomists wanted: that way, they could preserve the results of their experiments and show them to others at any time. This first became possible in Leiden in 1666 when Jan Swammerdam (1637–1680) developed solidifying fluid wax masses and thus successfully made injected preparations, which could be preserved either dried or in fluid.[17] Others in Europe were also experimenting with mercury injections and more permanent preparations at the time: Malpighi mentions using mercury to study the finest branches of the vessels of the lungs in 1661. The first account we have of a mercury injection in an experiment in the Netherlands is from Amsterdam in 1672, and Swammerdam was involved in it; it is therefore likely that he first experimented with hardening mercury injections in the same period.[18]

Within decades, the practice of hardening injections with mercury and coloured wax in anatomical preparations would spread throughout Europe, with varying success. In 1695, Blankaart appended a section on 'The new method of embalming' to a new Latin edition of his *Reformed Anatomy*. In it, he briefly mentions the possibility of injecting vessels with a mercury amalgam, but he seems unimpressed with the use of the technique for clarifying anatomical structures, stating that this was mainly done 'for financial gain.'[19] This remark suggests that at least to some late seventeenth-century anatomists mercury-injected anatomical preparations were merely commercial commodities, not research objects or scientific commodities that could be used in acquiring and exchanging anatomical knowledge.

Yet in *The Anatomy of the Brain* of the same year, Humphrey Ridley (?–1708), an MD from Nottingham who studied in Oxford, Cambridge, and Leiden, notes that an injection with mercury "... by its permanent nature and colour, contributes mightily towards bringing to view the most minute ramifications of vessels, and secretest recesses of Nature." Ridley preferred mercury to wax, the

17 Cole, "The History of Anatomical Injections," 301: Boerhaave mentioned 1667, but Swammerdam himself stated that he demonstrated his method of the solidifying injection mass to Van Horne, Slade, Thévenot and Steno in Leiden.

18 Lindeboom (ed.), *Observationes anatomicæ*, 23, *pars altera*, 15–16.

19 Blankaart, *Anatomia Reformata*, 1695, 758, XXX.

latter being too coarse for the finest vessels.[20] By the early eighteenth century, the Leiden professor and instrument designer Musschenbroek even offered a standardized instrument for mercury preparations: 'An Iron one [pipe] for Injecting of Mercury,' that cost 1 Florin, which would be about 10 Euros now.[21] But when young Nuck was coming of age and started to study medicine and anatomy in the 1670s, creating lasting anatomical preparations injected with waxes and mercury in order to discover and map bodily structures was still a new and exciting field.

Besides the fact that lasting preparations were difficult to make, their quality by the late seventeenth century was generally rather low. Dried preparations tended to shrink and discolour from their protecting layer of lacquer and were prone to damage by insects, larvae, and moulds, whereas wet preparations in alcohol suffered from discolouration, putrefaction and dissembling—the Ruysch and Albinus preparations are a notable exception to this. This also shows from the preparations kept in the Leiden anatomical theatre, established in 1594. According to Sandifort senior, in the first fifty years of its existence, the anatomical theatre was equipped with "the bones—either loose or kept together by their own connective tissue, or artificially reconnected and assembled—the skin and other parts, dried or put in alcohol without the right preparation" of those punished by military or civil authorities and dissected in the theatre.[22]

However, by the mid-seventeenth century, these preparations were "dilapidated by old age, or demolished by inconsiderate hands."[23] They were replaced when Louis de Bils, a Flemish nobleman well versed in anatomy, donated a number of preparations to the theatre in 1655 that were remarkably well preserved for the time. The most outstanding piece was a dried male human body, a stuffed skin complete with beard, scalp and eyes.[24] By 1771, the De Bils preparations would be discarded too, as they had either gone bad or had been badly damaged.[25] Although the dried preparations made by De Bils lasted already

20 Ridley, *The Anatomy of the Brain*, 4.

21 Erndl, *The Relation of a Journey*, 62.

22 Sandifort, *Museum Anatomicum*, XIIII: "Verum quod attinet ad praeparata Anatomica, solitum tantummodo fuit eorum, qui a politico vel militari Magistratu morte erant puniti, & in Theatro dissecti, ossa vel soluta, vel suis ligamentis adhuc cohaerenta, vel arte iterum nexa & composita, pelles, partesque alias, siccatas, aut sine praeparatione idonea liquori immissas ..."

23 De Bils, *Vertooch*, 7: "... door ouderdom vervallen, ofte door onachtzame handen gebrooken zijn."

24 Ibid., Huisman, *The Finger of God*, 85.

25 Sandifort, *Museum Anatomicum*, XIV.

twice as long as the earliest ones, their 115 years are child's play compared to the dried and wet preparations made in the late seventeenth and eighteenth centuries, still with us over three hundred years later. For these long-lasting preparations, mercury has a particular appeal: unlike coloured wax, it can even penetrate the smallest of vessels, it has the looks of a precious metal, yet it is cheaper and easier to inject than precious metals, and unlike other non-precious metals, it is not susceptible to corrosion and hence suitable for wet preservation too.

Moreover, the mapping of the lymphatic system and its function was a topical affair in the second half of the seventeenth century, a quickly developing field. Alexandrian anatomists Erasistrates and Herophiles both observed lymphatic vessels in animals in the third century BC, but thought them to be either arteries or veins. In the sixteenth century, the first accurate descriptions of lymph vessels appeared, most notably Eustachius' account of the thoracic duct (1564), the biggest lymphatic vessel in the body. It was only in the seventeenth century that it was established that the lymphatic vessels and glands are actually a separate and interconnected system, and even then its exact functioning and layout remained largely mysterious. Ruysch contributed considerably to the knowledge of the lymphatic system with his discovery of the valves in the lymphatic vessels, published in 1665, to which I will return later.

It would be Paolo Mascagni (1755–1815), a contemporary of Sandifort, who discovered in the 1780's about fifty percent of all the lymphatic vessels known today. Mascagni can also be credited with the finding that every lymph vessel enters at least one lymph node, and the disproval of the existence of arterial and venous lymphatic vessels. The exact function of the lymphatic system would remain a controversial topic up to the early twentieth century, not least because it is extremely difficult to discern the lymphatic glands in a dead body.[26] Their obscurity made it very complicated to devise the lay out of the entire lymphatic system, let alone to visualize it in an anatomical preparation. Mercury, whether or not mixed with other substances, turned out to be the perfect injection material for the lymphatics because of its penetrating qualities, as we will also see in the next section.

The possibility to create lasting preparations that showed the course of the lymphatic glands was therefore a welcome novelty in the late seventeenth century. However, if creating visually clear, lasting preparations was already problematic, making one which distinguishes the small and generally well-hidden

---

26 Ruysch, *Dilucidatio Valvularum*, Luyendijk-Elshout, "Introduction," Mascagni, *Prodrome D'un Ouvrage, Vasorum Lymphaticorum Corporis Humani*, Olry and Motomiya, *Paolo Mascagni*, Eales, "The History of the Lymphatic System."

lymphatic glands was even more difficult. Although many anatomists, including Nuck, were not very vocal about the problems and challenges they encountered when trying to make lasting preparations, there is a number of accounts from various periods that describe the difficulties of making mercury-injected preparations of lymphatic, and other, vessels. Some stress the importance of finding a suitable corpse to make a lymphatic gland preparation. Nuck's student Boerhaave's description of Nuck's techniques of preparing the lymphatic system mentions that a skinny corpse was preferred.[27] This preference had to do with the fact that the lymphatic glands tend to be covered with body fat in more corpulent bodies, something also mentioned by the Scottish anatomist Charles Bell over a century later, who wrote: "When there is much fat the dissection of them [the lymph glands] is difficult, and to preserve them it is absolutely necessary that the subject shall be thin, and anasarcous."[28]

The above indicates that the lymph glands are notoriously difficult to inject—according to Bell injecting the lymphatic glands was the most difficult part of practical anatomy.[29] Not only are the lymphatic vessels tiny and covered by other structures, the smaller ones cannot be filled through the larger ones (like blood vessels), because of the many valves in the vessels. Frederik Ruysch first described these lymphatic valves in *Dilucidatio valvularum in vasis lymphaticis et lacteis*, which he published shortly after obtaining his medical degree in 1665. In order to make the lymph glands visible in much wanted lasting preparations, a substance with special visual, penetrating, and lasting qualities was needed, and that was mercury. The importance of experiment and practice, of sensory and bodily involvement of the anatomist in making mercury-injected preparations was even greater than with wax-injected preparations. Because of the difficulty of the procedure and the spectacular results of successfully injecting mercury, it remained the ultimate proof of anatomical skill for almost two centuries. It was the combination of this somatic involvement with the aesthetic, symbolic, transmutational, and medical qualities of mercury that made it the perfect candidate for the exploration and mapping of the lymphatic system in the late seventeenth century, yet some of those selfsame qualities made it a controversial material for part of the eighteenth century.

---

27 Boerhaave, *Praelectiones Academicae*, 571.

28 Tompsett, *Anatomical Techniques*, 10–13. 'Anasarcous' means affected by dropsy (oedema), which causes the lymph glands to swell.

29 Ibid.

The Material Meanings of Mercury

The materiality of mercury in the Leiden preparations gives us access to, and is better understood through, a complex world of anatomical and medical knowledge and practices. These objects are, after all, evidence of complex social relationships; they are simultaneously products and modellers of a distinct epistemic culture.[30] Returning to the Klein and Spary statement that materials can be challenging things that speak irresistibly, not only through their interpretation and representation, we will see that mercury's agency and meaning are in its colour, its effect on the human body, and its volatile and intangible character.[31] This section will show that the changing uses and meanings of mercury in chemistry, medicine and specifically as injection mass from the seventeenth century onwards, are all closely connected and rooted in mercury's distinct materiality, and thus indispensable for understanding the Sandifort preparations.

Mercury was the most likely candidate for injecting the lymphatic vessels to seventeenth- and eighteenth-century anatomists for a number of reasons. There are practical material qualities that make it very suitable as injection mass, but as we will see it also had less obvious, symbolical properties in favour of it. Of course mercury's common name, quicksilver, immediately gives away the first material quality: its extreme mobility. Then there is the fact that even a very insignificant thread of it can be fairly easily discerned by the human eye, and that its silvery shimmer warrants a refined and pretty finish.[32] This quality ensures the use of mercury in preparations is compatible with aesthesis: it brings out a certain beauty in them. But what makes the materiality of mercury particularly significant in the case of the Leiden preparations is the changing meaning mercury had in medicine and chemistry in the long eighteenth century. Although often interrelated, for the sake of clarity the chemical and the medicinal meanings and uses of mercury in this period will be discussed separately here. Finally, the influence of these uses and meanings on the use of mercury as an injection mass in anatomical preparations will be illuminated.

30 Harvey, "Introduction: Practical Matters," 5, 12.

31 Klein and Spary, "Introduction: Why Materials?," 9.

32 Also see Tompsett, *Anatomical Techniques*, 10.

Chemical Mercury

To understand the changes in the materiality of mercury, we have to explore what mercury represented before 1700 and afterwards. As Pamela H. Smith has shown in her 2010 chapter on vermillion, mercury, blood, and lizards, early modern artisans understood the flow of blood and the flow of metal in similar ways, and sixteenth-century metalworking was part of a web in which vermillion, red, blood, mercury, gold, and lizards gave access to powers of nature, transformation, and generation. Mercury was traditionally associated with resurrection and transmutation.[33] In early modern alchemy, of which metallurgy was an important part, mercury played an essential role, together with sulphur. The idea that sulphur and mercury were the basic elements of all metals came from the works of Jabir Ibn Hayyan, which were translated from Arabic to Latin in the twelfth century.[34] Mercury was one half of the metaphorical chemical marriage between the royal couple of the red king (sulphur) and the white queen (mercury). These were principally philosophical concepts of sulphur and mercury, not identical to the material form of the two elements, but they drew on their physical characteristics.[35] Some alchemists even believed the copulation between philosophical sulphur and mercury produced the Philosopher's Stone, although how exactly that conjunction took place was far from straightforward.[36] The two elements of this metaphorical chemical marriage were opposites: sulphur was associated with dryness, fire and heat, whereas mercury was thought to be cool and wet, even to the extent that it was said to be found mostly in green, wet areas.[37]

But how were these traditional chemical meanings of mercury significant for the late seventeenth- and early eighteenth-century Leiden anatomists, and for Anthony Nuck, the first known Leiden creator of mercury-injected preparations, in particular? That can only be understood through the epistemological transformations in the field that Newman has termed chymistry—the totality of chemical and alchemical technology and theory in Early Modern Europe.[38] Within this totality, alchemy denotes the chemistry of metals, and transmutational alchemy, or chrysopoeia, the transmutation of metals. These

33 Smith, "Vermillion, Mercury, Blood and Lizards," 44–5, 47–8.
34 Ibid., 39.
35 Ibid., p. 40.
36 Kavey, "Mercury Falling," 127.
37 Smith, "Vermillion, Mercury, Blood and Lizards," 47.
38 Ibid., 499.

transformations, in full swing exactly when Nuck made his preparations, have been greatly clarified in recent historiographical work by Newman and Principe. Chemistry was, after all, not a discipline within the early modern university, although it was certainly also practiced by university-educated men. Concerned not only with the quest for the Philosopher's Stone, chrysopoeia (metallic transmutation; particularly making gold), but also with artisanal practices such as making pigments and medications, chemical practice was long situated mainly in monasteries, courtly, and medical circles.[39]

Alchemists were alternately viewed as either wise men or quacks and frauds, as is illustrated in two seventeenth-century paintings of alchemists: one from 1661 by Adriaen van Ostade (1610–1685), and one by a follower of David Teniers the younger (1610–1690).[40] Whereas Van Ostade's alchemist is a laughable figure, burning away his last possessions in a messy workshop, the alchemist painted by Teniers' follower is a respectable intellectual, a man of the world, (symbolized by the globe), consulting a book in a fairly organized studio workshop.[41] However, these contrasting and sometimes outright negative portrayals started to change in the course of the seventeenth century, when the alchemical emphasis on experimental analysis became increasingly important in natural philosophy and medicine. Methods for, and knowledge about, the purification of metals turned out to be indispensable for seventeenth- and eighteenth-century medicine and chemistry.[42]

For example, in the 1623 catalogue of the Leiden anatomical theatre, a coloured print of Breughel the Elder's *Alchemist* is listed, an image in which the alchemist is depicted as a money-wasting quack, destined for the poorhouse.[43] However, in 1669, the first chemical laboratory at Leiden University was estab-

39 Newman, "From Alchemy to 'Chymistry'," 498.

40 The painting by Van Ostade is at the National Gallery in London, see http://www.nationalgallery.org.uk/paintings/NG846. The painting by the follower of David Teniers, *An alchemist in his laboratory*, is in the iconographic collections of the Wellcome Library, accession number L0051290, and is accessible via Wellcome Images, http://wellcomeimages.org.

41 Newman and Principe, "Alchemy Vs. Chemistry," 38: Early modern humanists such as Erasmus and Petrarch even displayed a general disdain for alchemy, but others, like Agricola, recognized its potential in natural philosophy. The latter tried to 'purify' the term alchemy to its Greek root *chymeia* (probably derived from the word for smelting metals, *cheein*) by discarding the Arabic definite article *al*. This resulted in the sixteenth and seventeenth century in the alternate use of forms of both words, *alchemia* and *chemia*, sometimes even in the same text, without significant distinction in their meaning.

42 Roberts, "Death of the Sensuous Chemist," Powers, *Inventing Chemistry*, 141–169.

43 Barge, *De oudste inventaris*, 43.

lished, "to complement the medical faculty" and where the students could be trained in the "operations and experiences of Chemistry."[44] By the late seventeenth century, most natural philosophers used theories and experiments that originated in alchemy, but they wanted to distance themselves from certain associations alchemy also had with deceptive and secretive arts like magic and witchcraft.[45] Transmutational alchemy, the branch of alchemy concerned with chrysopoeia, was increasingly vilified by natural philosophers worried about their social status and reputation.[46] By narrowing their definition of alchemy to metallic transmutation, and defining chemistry as a new and respectable discipline, they aimed to incorporate it into academia, and indeed did so with eventual success. Newman has situated the real divorce between 'alchemy' and 'chemistry' in Lemery's conscious excision of 'alchimie' (metallic transmutation) from the 1679 third edition of his popular *Cours de Chimie*, and, by the first decades of the eighteenth century, natural philosophers were widely ridiculing 'gold-making' and portraying alchemists as frauds.[47] Yet chemistry was not incorporated into academic discourse without a struggle, and some of the chemists who publicly rejected chrysopoeia continued to pursue it privately, while others maintained that metallic transmutation was just a subset of chemistry, and as late as 1783 a claim that gold could be made through metallic transmutation was seriously investigated by the Royal Society.[48]

These tendencies were noticeable within the Leiden academy as well. In his inaugural lecture of 1731, Gaub pleads for the incorporation of Chemia in the academic arts. In his address, he stressed that although chemistry, with its

44 Resolutions of the Curators, 8 August 1668, quoted in Jorissen, *Het Chemisch Laboratorium*, 14: "Naedemale tot perfectie der medische faculteijt in deze Universiteijt noch ontbreeckt een laboratorium Chijmicum, waerdoor de studenten in operationibus et experiential Chijmica konnen geoeffent werden; soo is bij den Heeren dese vegaderinge aangenomen, iemant te dispicieren, dewelcke een soodanigen laboratorium Chijmicum sal connen oprechten, ende also de ervarentheijt der natuerlicke operatien door het middel der Chijmie te demonstreren."

45 Moran, "Introduction," 302.

46 Principe, "Alchemy Restored," 312.

47 Knoeff, *Herman Boerhaave*, 14–16, Newman, "From Alchemy to 'Chymistry'," 511, Newman and Principe, "Alchemy Vs. Chemistry," 39. Interestingly, art historian De Clippel, "Alchemist / Alghe Mist," 35–7, also identifies the increasing incorporation of alchemy in natural philosophy in iconography in the Low Countries: around 1670, depictions of 'alchemist's workshops' increasingly show studies filled with books and less laboratory-like settings with instruments than before.

48 Principe, "Alchemy Restored," 306, Neumann, *Medicinale En Natuurkundige Chymie*, 8, Fara, "Marginalized Practices," 485.

dirty hands-on experimental practice, seems far removed from the elegance of rhetoric, a visit to a chemical workshop ('officinam Chemicam') would make clear that chemistry does not delight the ears but the eyes, that here it is not words but demonstration that delivers facts. Gaub argued that philosophy and medicine are rightfully academic disciplines because they know and cure the body; chemistry can contribute to this by providing knowledge about the constitution of the material of bodies.[49] Seven years later, Abraham Kaau gave a declamation in Leiden satirizing alchemy, portraying the alchemist as greedy, secretive, stubbornly foolish, and diametrically opposed to the systematically working chymist, who aimed to acquire knowledge about the material world and did so openly.[50]

In Leiden, the public defence of chemistry and ridicule of alchemy were the accumulation of a process that had started well over twenty years earlier, when Kaau's uncle, the Leiden professor Herman Boerhaave (1668–1738), did not mock chemistry's history, but apologized for it, as he was embarrassed by what he thought to be its excrescences (greed, silliness), but who also was excited about what it could mean for medicine and natural philosophy.[51] Therefore he was also propagating the new chemistry, in which mercury was still an important element. Boerhaave initially believed quicksilver to be a fixing principle of metallic bodies, without which all metals would be loosely floating particles. It was through studying elements like mercury that he could glimpse the perfection of the Creator. After all, the symbols for mercury and gold suggested that gold was a purified form of mercury—the purest metal actually possible.[52]

Boerhaave performed many experiments attempting to purify mercury, and through this chemical experimentation came to the conclusion that no mercury was 'fixed' into gold or silver, regardless of how long it stayed in the fire, how strong the fire was, or how many times it was distilled. It was through these experiments that he eventually changed his mind about the possibility of metallic transmutation.[53] Boerhaave stated that "it does not appear through these experiments that mercury and fire may form metals. Therefore fire … is not demonstrated to be the Sulphur of the Philosophers, fixing mercury into metal."[54] This experiment and conclusion show that, although Boerhaave

49 Gaubius, "Inaugurele Rede," 177–199.

50 Kaau, "Declamatio Academia," 6.

51 Principe, "Alchemy Restored," 305–6, Boerhaave, "Sermo Academicus De Chemia."

52 Knoeff, *Herman Boerhaave*, 144–6, 211.

53 Ibid., 151.

54 Powers, "Scrutinizing the Alchemists," 235, Boerhaave, *Some Experiments Concerning Mercury*, 162.

and his contemporaries distanced themselves from transmutational alchemy, they were still familiar with the traditional alchemical knowledge of metals and materials. They took from it the emphasis on analysis and the quest for the reduction of materials to their pristine state, and used these approaches to investigate traditional claims about the properties of materials such as mercury.

Hence the supposed complicity of vermilion red and white, sulphur and mercury, fire and water, in addition to the centuries-old association of blood and mercury, was still widely known by the late seventeenth and early eighteenth century. Although nowhere mentioned explicitly, these associations may well have been a reason for anatomists to choose vermillion red wax and silvery-white mercury to inject, respectively, the veins and lymphatic glands in anatomical preparations. It gave those preparations an additional elegance: not only do they look elegant, but their materiality may well have referred to old chymical and medical knowledge too. After all, according to the predominant Hippocratic humoral theory, the red blood was concocted (literally 'cooked') in the liver, it nourished the organs and was associated with hotness and dryness, whereas the function of the lymphatic system was largely a mystery, but with its mostly invisible contents and the milky fluid found in the chyle it might well have be the wet, cold, balancing opponent of the blood. What better way to show this than with red wax for blood and silvery mercury for the lymphatics?[55]

For the medical men and anatomists occupied with exploring the lymphatic system and circulation of the blood, and the making of lasting anatomical preparations in the second half of the seventeenth century, an anatomical preparation injected with red wax and solidified mercury would therefore have been quite easy to read and use: red wax for the hot, red blood, mercury to visualize the otherwise almost invisible cold, wet, whitish lymphatic vessels. In the early eighteenth century transmutational alchemy was increasingly condemned in favour of the supposedly more academic, but actually hard to distinguish, chemistry. It is therefore not unimaginable that, at least for some anatomists, mercury, with all its transmutational connotations, at certain moments appeared to be less desirable as an injection mass. In addition to the suitable symbolism mercury provided to the anatomist, it had another material

55 Luyendijk-Elshout, "Antony Nuck," 17–29: The chyle duct and its function were topics of heated debate in 1650s Leiden. Witkam, "Catalogues Anatomy Hall,": As appears from a description of the English 1712 catalogue of the Leiden anatomical theatre, mercury was incidentally also tried on preparations of other structures. On page 22 "The Matrix, and a stone of a Man spowtid with quikzilver." is listed, but this seems to be an exception.

quality that made it a very appropriate injection mass—a quality that also made it popular in medicinal use.

Medicinal Mercury

Apart from symbolical and philosophical meanings and metallurgic uses, mercury also had a long history as a remedy. Traditionally, mercury was used to cure skin eruptions and venereal diseases. Although its usefulness as a cure for syphilis was debated from the early eighteenth century onwards, for want of a more effective cure mercury continued to be prescribed for venereal diseases well into the nineteenth century. This ongoing use was rooted in the idea that the all-penetrating qualities of mercury would expel poisons and disease from the body.[56] Hence early eighteenth-century mechanistic professors of medicine and chemistry in Leiden also praised mercury for its purifying qualities: its particles were thought to cleanse even the tiniest canals of the body from misbalancing impurities.[57]

However, by the mid-eighteenth century, it was already widely known that mercury could be dangerous too when used imprudently. For example, in the 1740 *Medicina Pharmaceutica*, published by the Leiden apothecary Johannes Schróder with a foreword by professor Gaub, mercury is said to induce drooling and is listed in different concoctions as a cure for "many vicious diseases," but the author also warns that it can be extremely dangerous when used in the wrong way.[58] And in Houttuyn's 1761 book on natural history it is noted that a dog that licks the saliva of someone who started drooling from a mercurial cure may die.[59] Yet mercury, either pure or combined with other substances, remained a standard ingredient in medication for decades to come. Gerard van Swieten (1770–1772) had studied with Albinus in Leiden in the 1720s and later became the personal physician of Maria Theresia of Austria; he is known to have experimented with making sublimated or purified quicksilver, and to have avidly prescribed it to syphilis sufferers during his years in Vienna. He also praised the accuracy of mercury-filled thermoscopes (thermometers), first made in 1724 by D.G. Fahrenheit.[60] In the 1805 *Pharmacopea Batava*, commissioned

56 Swiderski, *Quicksilver*, 7, 97.

57 I.e. Nuck, *Adenographia Et Uteri*, 37: "Mercurius namque vel maxime exiguos glandularum canaliculos in tantum aperit, ut excretio fiat satis copiosa . . ".

58 Schroder, *Schatkamer der Droogbereidende Geneeskonst*, vol. 1, 222, vol. 3, 153–4.

59 Houttuyn, *Natuurlyke Historie*, 1761, 67. Also see Dewhurst, *The Quicksilver Doctor*.

60 Van der Korst, *Een Dokter Van Formaat*, 40, 173.

by the government and compiled by a committee led by the Leiden professor of anatomy Sebald Justinus Brugmans, a full six pages are still devoted to this substance which consists of "shiny, exactly spherical balls running everywhere," and which "in the fire calmly and completely evaporates."[61]

Mercury as Injection Mass

As we have now seen, the materiality of mercury as an (al)chemical element and as a drug with particular material, symbolical, and visual qualities, cannot be considered separately from its use as an injection mass in anatomical preparations by the late seventeenth century. Previously, historians have suggested that mercury ended up in the hands of anatomists after having first been used by their suppliers, instrument makers, and in experimental baro- and thermometers, but in the past decades it has become clear that modern professional categories fail to describe and understand early modern ways of knowing.[62] Men like Ruysch, De Graaf, Swammerdam, Nuck and even Boerhaave were polymaths; all were (al)chemist, instrument maker, anatomist, physiologist, apothecary and medical doctor at the same time. Nuck praised mercury for its penetrating qualities as follows in 1685, in a work on the tear and salivary ducts:

> Mercury opens the small canals of the glands so widely that secretion can occur abundantly. Once mercury has made itself part of the blood, divided into an innumerable quantity of small round particles, it easily adapts itself to every shape of the pores and shatters the sharp deposits of salts where they have settled. When they have been brought into motion, most of the obstructions are abolished, and when these are removed all the disease-generating substances are washed away through the salivary ducts.[63]

---

61 Brugmans et al., *Pharmacopoea Batava*, 22, 179–85. Mercurial compounds would be used in drugs until the 1930s, see Swiderski, *Quicksilver*, 119, 141.

62 Luyendijk-Elshout, "Antony Nuck," 156: suggests late 17th C. 'research physiologists' were introduced to mercury through the first baro- and thermometers, an idea refuted by the more recent works of i.e. Newman, Principe, Gallison and Daston.

63 Nuck, *De Ductu Salivali Novo*, 37, translation from Luyendijk-Elshout, "Antony Nuck," 159–60.

Because of these qualities, mercury was thus both very suitable as the main ingredient of a penetrative injection mass, and as a cleansing medicine, that removed all kinds of obstructions. Apart from these curing qualities there were more allegorical meanings inextricably attached to quicksilver: it was traditionally associated with resurrection, as it separates into innumerable tiny, perfectly round balls but transforms back into a smooth mass when these are captured in a container.[64] And then there is of course the god Mercury, protector of merchants and thieves. Nuck repeatedly referred to mercury as 'Noster Mercurius,' which Luyendijk-Elshout translated as 'friend mercury.' Yet written with capitals, this term refers not to the common metal, but to philosophical Mercury, a union between sulphur and argent vive or first mercury.[65] Moreover, to explain his work on the lymphatic system, Nuck employed an elaborate analogy in his 1691 *Adenographia*. He compared himself to a Mercator, a seafaring merchant, who sails unexplored rivers of bodily fluids to discover unknown lands and treasures.[66] In addition, the planet Mercury was traditionally thought to influence the hands, the anatomists' most important instruments together with the eyes, and to be "a bright star, a great source of eloquence, of subtle ingeniousness and of the fine arts, and the enemy of vanity."[67] From this it appears that Nuck's reasons to try and use mercury as an injection mass were multiple and not all purely practical.

This quickly becomes clear when the disadvantages of using mercury as an injection mass are considered, and which Nuck and his contemporaries must have been confronted with as well. Quicksilver is heavy, and if a column of it builds up inside a vessel, it easily ruptured the vessel, which happened frequently if slightly too much was injected. Even if an injection was successful, the resulting preparation would remain very delicate, especially if the mercury mass had not dried yet, and even careful handling could destroy the preparation. As noted before, injecting anatomical structures with mercury was thus something that not only required the right equipment, practice, a sharp eye for detail, and patience, but also tacit bodily knowledge, a practical skill that is hard, if not impossible, to describe in words. Just as he had taught Boerhaave, Nuck knew that the early stages of decay or moderate oedema made the lymphatic glands easier to find. To fill the capillary lymphatic vessels ('saugadern'),

64 Abraham, *A Dictionary*, 125, Dobbs, *Alchemical Death and Resurrection*, 21–22.

65 Abraham, *A Dictionary*, 124.

66 Nuck, *Adenographia Et Uteri*, 4.

67 Bussagli, *Het menselijk lichaam*, 216, quotes from a fifteenth century Italian print of the spheres of Mercury: "Mercurius di ragion lucida stella/produce d'eloquenza gran Fontana/ subtili ingegni et chiaschun arte bella/ et è nimico d'ogni casa vana."

Nuck chose a very small vessel and inserted a tiny iron tube through which the mercury, mixed with a hardening agent such as tin or lead, was carefully injected.[68] However, these kinds of descriptions would be of little help for someone wanting to make their own mercury-injected preparations. This brings us to an important aspect of aesthesis: tacitness. Creating a mercury-injected preparation is largely tacit bodily knowledge, something that can only be learned through endless practice and experience.

In general, seventeenth- and eighteenth-century how-to descriptions of injecting preparations, with mercury or otherwise, are few and far between. The descriptions that can be found tell us nothing about how much mercury should first be used, in which proportions to mix the mercury with other metals, and at which temperature to keep it in order for it to remain fluid during the process of injecting without scorching the preparation and allowing it to harden once inside the vessels. Nor do they say anything about exactly what instruments to use, how much pressure should be applied, or how the finished preparations should be best handled. Many of these details may not have been consciously known by anatomists such as Nuck themselves, as they developed the technique by actually doing it. Even if the technical details about tools, mixtures, and temperatures would have been written down, simply just following these descriptions would not have guaranteed a successful mercury-injected preparation. Even Thomas Pole, writer of an exceptionally detailed 1790 handbook on dissecting and making anatomical preparations, warned the aspiring anatomist that

> In making quicksilver Injections, the principal ingredients, and the first to be obtained, are time and patience, and not less so, an uniform fortitude against disappointments; for it will not unfrequently happen, that with the greatest care, a most promising preparation will be instantaneously destroyed by some trivial accident, when it has been almost completed.[69]

As Pole also pointed out, a handbook of anatomical techniques could never be more than a guide; the only way to learn anatomy was through doing it.[70] The experimental character, the lack of description, and the difficulty of sufficiently describing such a procedure, combined with the notorious difficulty of injecting the lymph glands and the delicacy of the desired result, all stress the somatic tacitness of injection practices. Somatic tacit knowledge, a term

68 Schultka and Göbbel, "Präparationstechniken Und Präparate," 72–3.

69 Pole, *The Anatomical Instructor*, 60.

70 Ibid.

borrowed from Polanski and Collins, describes something which can be explained in theory, but the only way to learn it yourself is to bodily involve yourself in it. In that sense, injecting anatomical preparations is the kind of knowledge comparable to riding a bicycle: impossible to learn without practice.

The combination of symbolical, chymical, and medical meanings and uses of mercury, and the bodily involvement required to learn how to inject mercury mixes and create lasting preparations, fits very well with the combination of seeking beauty and the importance of sensory perception and materiality in aesthesis. Mercury as a material ensured that anatomical preparations were elegant, refined artefacts. Especially for the initial makers of such preparations, such as Nuck, mercury-made preparations refer to a centuries-old body of the mystical and practical meanings of the substance, such as resurrection and purification, exploration and refined ingeniousness, as well as the possible structure and function of the lymphatic system that they demonstrated. The well-known impalpable character of mercury meant that the acquiring of the kind of somatic tacit knowledge needed to make a mercury-injected preparation became a way for an anatomist to prove his worth. In developing these skills, he showed he was well-versed in natural philosophy and chemistry, that he was able to use his sensory perceptions to gain knowledge of the body, and, to spread that knowledge in a beautiful, refined and perfect manner, and, last but not least, that he himself had also reached a level of perfection. Although the more symbolic and transmutational meanings of mercury became outmoded in the course of the eighteenth century, the material qualities of quicksilver meant it was never abandoned altogether as an injection mass for anatomical preparations. An anatomist who knew his injection techniques, especially those involving a material as hard to handle as mercury, was an elegant anatomist. This would prove to remain an objective for many years to come, although the means to reaching that goal were subject to change.

Mercury and Aesthesis in Leiden Anatomy between 1700 and 1771

What role, then, did mercury play in aesthesis, and the anatomical practices and artefacts that were a part of it, in eighteenth-century Leiden? As mentioned before, Boerhaave in his lectures spoke admiringly of Nuck's mercury-injected, dried preparations of the lymph glands, and performed chemical experiments to establish the properties of mercury. As we have seen in the previous section, Nuck was familiar with mercury and its uses—not altogether surprising considering he started his career as an apothecary; his successors did not shy from using it as a drug either. In a 1740 reprint of his 1692 manual of surgery, cures, and healing for example, mercury is still listed as an ingredient of a concoction to remove sputa in the nose.[71] It has even been suggested that Nuck's rather excessive use of his 'friend mercury' might have been the cause of his death after a short but severe illness.[72] However, another, more contemporary, source states that while dissecting a dead body, Nuck cut open a lesion that was filled with a "foul and malicious" matter, which attacked his heart and swiftly "cut his life cord."[73] Hence after his death in 1692, Nuck's colleagues and successors in Leiden and abroad initially continued to use mercury in preparations in their quest for knowledge, perfection, and elegance.

Few mercury-injected preparations from around the turn of the century survive however. This may have several reasons, such as the fact that even a finished mercury preparation is very fragile and thus easily damaged, but also could be due to the fact that the preparations are so difficult to make; the high price of mercury may have also contributed to this scarceness.[74] Nuck's contemporary Ruysch, for example, used mercury injections rather sparingly in his many preparations, but he certainly appreciated its material and philosophical qualities. He wrote a treatise on the valves in the lymph glands when he was still a student (he graduated in 1664). Ruysch prepared his book by doing experiments in the years before 1665 to settle a long-running dispute among Leiden faculty and students.[75] He suspected there were valves in the lymph glands because he felt resistance when he was probing them with a small copper tube. In order to make them visible, he blew air into what he called the 'water vessels,' and then dried his preparation in the sun and the wind. He then 'saw the little valves, appearing like half-moons.'[76] Later in his career, Ruysch would use mercury to inject the smallest vessels, and there were allegorical references to mercury in his cabinets, like children's skeletons holding serpents that bit their own tail.[77]

---

71 Nuck, *Genees- Ofte Heelkundige Handwerken*, 56.

72 Luyendijk-Elshout, "Antony Nuck," 158.

73 Overschuur in Nuck, *Genees- Ofte Heelkundige Handwerken*, preface: "...door het Ontledigen eenes Doode-Lichaemes, waer in hy [ô droevig ongeluk!] een zekere blaesie kwam te openen, vervuld met een zeer kwaedaerdige en fenynige stoffe, die hem vervolgens nae het herte sloeg, den draedt zyns levens schielyk is afgesneden geworden....".

74 Chaplin, "John Hunter," 115, 143.

75 Luyendijk-Elshout, "Antony Nuck," 34–9.

76 Ruysch, *Dilucidatio Valvularum*, 3–4.

77 Abraham, *A Dictionary*, 181: "The uroboros or paradoxical serpent, which devours its own tail and begets itself, is a symbol of the circular process of the opus alchymicum." Edward Kelly wrote of Mercurius: "*It is the wanton serpent that conceives of its own seed, and brings forth on the same day.*"

Unfortunately, only one mercury-injected preparation made by Ruysch is known today. This preparation is now housed in the Kunstkamera, and Ruysch describes it as "Some membrane of which the arteries are filled with living mercury," suggesting that in this case he had tried mercury as an enlivening injection mass.[78] There is a fairly straightforward explanation for Ruysch's sporadic use of mercury as an injection mass. He probably found the silvery shimmer simply unsuitable for many of his preparations, as his objective was often to make them look as life-like as possible—something that could not be attained by using mercury. In Leiden too, the habit of making and collecting mercury-injected preparations established by Nuck continued for some time—although it did not seem to last.

Sometime between 1701 and 1707 the collections of Govert Bidloo (1649–1713), professor of anatomy at Leiden University from 1694 to 1713, were added to those of the Leiden anatomical theatre, and the 1707 visitor's catalogue is the first to mention a number of preparations kept in liquor, as well as preparations injected with mercury and coloured wax.[79] No makers are listed, but the mercury-and-coloured-wax preparations could also be part of Nuck's legacy. In the decade that followed, few preparations were added to the collection in the anatomical theatre, most of them curiosities such as shoes made of a man's skin and acquired by the keeper of the theatre, Gerard Blancken.[80] Only with the death of professor Johannes Jacobus Rau in 1719 was a substantive addition made again. Rau stressed the importance of gaining knowledge about the body through dissection, and making injected preparations, throughout his appointment, starting with his oration in 1713.[81] However, he was weary of the kind of preparations his colleague and life-long competitor Ruysch made. In his oration he warned his students of a lack of naturalism in preparations in which artefacts were used, and Erndl, a Saxon physician who travelled through the Netherlands in 1707 and took classes with Rau, noted that when the students asked Rau his opinion of the preparation of the entire body of a boy of about eight years old kept by Ruysch:

> Mr. Raw [Rau] has said often, when we expected his Opinion concerning this Prodigy; that there was some wicked Arts us'd with the Boy, other-

---

78 Ruysch *Thesauri*, vol. 9, 1726, 39, no. 94: "Membrana Quaedam, cujus arteriae mercurio vivo sunt oppletae."

79 Witkam, "Catalogues Anatomy Hall,": 1707.

80 Witkam, "Catalogues Anatomy Hall,": 1707, 1709, 1712, 1719.

81 Rau, *De Methodo Anatomen Docendi.*

> wise the Skin with the Muscles cou'd never have been so full of Blood, and kept their natural Colour.[82]

In addition, when Uffenbach visited Rau's collection in his Amsterdam home in 1709, he remarked that many phials of the preparations were only half-filled with liquor—Uffenbach suspected Rau was too stingy to spend money on replenishing them.[83] It is therefore not surprising that Rau's inheritance seems to have consisted mainly of instruments for cutting stones and bone preparations—Rau's specialities—and some preparations of the sensory and reproductive organs.[84] No mercury preparations were found in Rau's collection. In the 1720s, Bernard Siegfried Albinus, the young new professor of anatomy, a former student of Ruysch and Rau, catalogued and organized the newly acquired collections. Meanwhile, he started working on his own preparations, which would later become the showpiece of the Leiden anatomical collections. In Albinus' work, the influence of his teachers was clearly visible. As we will see, his preparations in which lace sleeves and pieces of plants were used were clearly inspired by Ruysch, whereas his more austere osteogenetic preparations fit better within the tradition of Nuck and Rau.[85]

Mercury injections did not seem to play a part at all in Albinus' preparations: none of his preparations mentioned in Sandifort's Museum Anatomicum have been injected with mercury. The only two mercury-injected preparations listed under his collections are, as we have seen, ascribed to Nuck, albeit doubtfully.[86] One of Sandifort's students mentioned the use of a preparation of a mercury-injected testis during his lectures, but this seems to be an exception and might well be a legacy from either Nuck or Ruysch.[87] The almost complete absence of mercury preparations can be explained by the fact that Albinus' greatest research interests, such as osteogenesis, sensory perception, and the functioning and colour of the skin, required preparations in which mercury was of little use as an injection mass. As will be discussed in more detail in the next chapter, Albinus too strove to be an elegant anatomist, to reach perfection in all

---

82 Erndl, *The Relation of a Journey*, 65–6.

83 Von Uffenbach, *Merckwürdige Reisen*, vol. 3, 621.

84 Albinus, *Index Supellectilis Anatomicae*.

85 See chapters 3 and 6.

86 Sandifort, *Museum Anatomicum*, vol. 1, p. 89, describes no. 407–408 (Albiniana) as follows: "CCCCVII. Portio intestine cum portione mesenterii, vasis nonnullis mercurio impletis. Nuckii ut videtur. CCCCVIII. Portio intestine, cujus vasa lacteal impleta mercurio. A Nuckio, aut Swammerdammio, ut videtur." This suggests he was not entirely certain about the origins of these preparations either.

87 Box MS UvA, f596, hs II G24–27.

his work, but as his research interests were not the lymph glands, he chose to express his elegance, his skill, and quest for beauty in coloured wax preparations and allegorical suspensions instead of mercury preparations. In Leiden, the first mercury-injected preparations to become part of the university collections after those made by Nuck were therefore those that first appeared in catalogues of the collections in the second half of the nineteenth century, and appear to have been made by Eduard Sandifort.

Ruysch and Albinus showed little interest in using mercury as an injection mass, and between 1710 and 1760, we find no mentions of mercury-injected preparations in catalogues or elsewhere in the Leiden archives. Neither did anything new seem to happen in the field of lymphatic research. It appears the focus of anatomists had shifted to other topics, most notably the spirit of life or vis vitalis, and the sensibility and irritability of body tissue.[88] Was this the only reason no more mercury preparations seem to have been made for a while, or was it also because making mercury-injected preparations was so hard, so time-consuming and expensive? Or were those five decades of relative quiet caused by the dubious transmutational and medical properties ascribed to mercury? It is hard to tell, as none of the Leiden anatomists, or their students and visitors from the period, appears to have explicitly denounced the making of mercury-injected preparations for researching the lymphatic system, but the temporary halt in the creation of mercury-injected preparations was probably due to a combination of all of these factors.

However, from the 1760s onwards, lymphatic research once more became a major topic. From the 1750s, Alexander Monro II (1733–1817) and William Hunter (1718–1783) had been in a public dispute about who first discovered the lymphatic vessels to be absorbent.[89] In 1760, treatises by Alexander Monro II and J.F. Meckel appeared, challenging Nuck's views on the lymphatic system, and marked the start of renewed, international attention upon the lymphatic system. Nuck and his direct successors had held the view that the lymphatic system was interconnected with the arterial system. Yet Monro II had noticed that in cases of inflammation, blood tended to bypass the lymph glands and flow directly into the smallest vessels. Meckel, who had specialized himself in the nerve system, set out to repeat Nuck's *Adenographia*. He once again used

88 See i.e. Steinke, *Irritating Experiments*.

89 Ambrose, "The priority dispute." It was actually Francis Glisson who discovered this first, but his mid-seventeenth century publication on the lymphatic system did not resurface until Monro II and William Hunter had independently reached the same conclusion. Hunter stopped battling with Monro II the moment he realized that Glisson was actually the first to describe the absorbent properties of the lymphatic vessels.

mercury to trace the lymph vessels, studied the nodes closely, and came to the conclusion that the lymphatic system was in fact an entirely independent entity, with a flow of its own.[90]

The lymphatic system was back on the anatomical research agenda once again. The Italian Paolo Mascagni (1752–1815) also returned to the injecting techniques originally developed in the second half of the seventeenth century to study the lymphatic system. The result was an initial exploratory account published in 1784, followed by a magnificent atlas of the lymphatic system that appeared in 1787, and that would remain the authoritative guide on the subject for over a century.[91] In the preface, Mascagni notes that he was inspired by Nuck's mercury injections to trace the lymph vessels.[92] Mercury was the material of choice once again because its cleansing qualities fitted well with the newly discovered absorbent properties of the lymphatic system. This is precisely the period in which Eduard Sandifort also came of age, graduated, and was appointed professor at Leiden University.

Hence it is no surprise that there are so many anatomical collections from the second half of the eighteenth century that contain at least a few mercury-injected preparations, not only in the Leiden collections, but also elsewhere in Europe. For example, in the 1778 catalogue of the sale of the private collection of the London surgeon and professor of anatomy Magnus Falconar, a fair number of the wet and dried preparations listed had been injected. Most of them apparently only with red and yellow wax, but of the preparations of the lymph glands, 13 out of the 16 preparation mentioned are explicitly described as having the lymphatic vessels injected with mercury. Mostly in these examples the arteries are injected with 'red' and the veins with 'yellow.' This also goes for the Preparations of Glands: all four preparations are said to be injected with mercury. Another section on lymphatic glands lists 14 mercury-injected preparations.[93] John Hunter was apparently so proud of a mercury-injected epididymis (the coiled tube on the back of the testicle) of a boar in his collection that he had it mounted in a gilded, glass-covered frame.

The recurrent but incidental appearance of mercury-injected preparations, mainly of the lymph glands, in late eighteenth-century anatomical collections reconfirms a number of points. First, that this period saw a renewed interest in research on the structure and function of the lymphatic system. Second, that it was still very hard to create a proper mercury preparation: it required

90 Monro and Meckel, *Opuscula Anatomica*, Luyendijk-Elshout, "Antony Nuck," 163.
91 Mascagni, *Prodrome D'un Ouvrage, Vasorum Lymphaticorum Corporis Humani.*
92 Mascagni, *Vasorum Lymphaticorum Corporis Humani*, 36.
93 Paterson, *Museum Falconarium*, 13, 27.

a lot of skill and practice to create one in which the lymph glands were made visible without making them burst. But if it worked however, the result was stunningly beautiful. Put one mercury-injected preparation on a shelf amongst a host of other preparations, and your eyes are drawn to its shimmer even from a distance. Third, that mercury was seen as an appropriate injection mass for lymph glands only: veins and arteries could better be injected with coloured wax-based mixtures, but the almost invisible lymphatic glands with their mysterious contents and function were best marked by an equally intangible, still slightly mystical substance like mercury. These qualities explain why mercury and its properties suited the first Leiden investigator of the lymphatic system, Nuck, so well. Albinus, although an elegant anatomist in all respects, chose not to use mercury because of his interest in entirely different research topics. And, as we will see, the material meaning of mercury was also the reason Eduard Sandifort returned to this injection mass by the late eighteenth century.

The Meaning of Mercury to Eduard Sandifort

The eighteenth-century historiography of researching the lymphatic system, and the mercury-injected preparations resulting from it, show that it is very well possible that the old mercury preparations in the Leiden anatomical collection are indeed Eduard Sandifort's work: it was a hot topic in his time and field, he had the means and skills to make them, and based on catalogue records, the preparations must have entered the Leiden collection somewhere between 1770 and 1860. What remains curious though is that there is no mention of them during Sandifort's lifetime, or of any attribution of these preparations to his own handiwork by any of his close successors, including his own son. Sandifort may have used the mercury preparations in his lectures on anatomy and physiology, but unfortunately no lecture-notes remain.[94] What were the possible motives for Sandifort to make these preparations, and for what reasons did he and his successors feel it was unnecessary to explicitly mention them and attribute them to his person? In order to answer these questions, it is first necessary to look at the circumstances under which he arrived in Leiden in 1771.

94 Molhuysen, *Bronnen*, vol. VII, 1924, Bijlage, 84, 13 Maart 1807: "Prof. Ed. Sandifort heeft sedert den jare 1770 standvastig lessen gegeven: over alle deelen der Anatomie (zoo publiek als privaat), over de Chirurgie, over de Morbi ossium, over de Fundamenta Artis Obstetriciae, over de Physiologie." (Molhuysen also notes that Eduard's son Gerard has taken over the private anatomy lessons and the demonstrations now.)

In 1769, Eduard Sandifort was a Leiden University graduate and practising city physician in The Hague. When he was appointed as lector of anatomy at Leiden University in 1771, he faced a number of challenges. B.S. Albinus died in 1771, and he and his aesthesis had dominated Leiden anatomy teaching for years. Although Albinus' anatomical cabinet had long been housed on university premises he did not bequeath it to the university and it thus became the property of his wife upon his death. Undoubtedly aware of the fame and value of the collection, she decided to turn it into cash and auctioned it off as a whole, together with her late husband's personal library, which was sold in parts. This caused a great stir among anatomists and gentlemen natural philosophers both in the Low Countries and abroad. Leiden University managed to acquire the cabinet for 6,300 guilders, a sum that today would equal about € 55,500.[95]

In short, the Albinus cabinet, although skilfully crafted, was relatively old-fashioned with its lace-wrapped, perfect, non-pathologic specimens of hands and heads—not exactly what was needed for teaching in the ailing Leiden medical faculty. When Albinus had been appointed in the early 1720's, Leiden had been one of the outstanding centres in Europe for the study of medicine. However, fifty years later, with Boerhaave long dead, Albinus an almost blind old man, and Holland's glory as a seafaring nation in decline, student numbers faltered and the medical faculty was in dire need of modernisation. This also shows from several travel accounts. Poole, already in 1742, describes a visit to the university and laments Boerhaave's death, but does not even mention the anatomical theatre or collections, and also seems to be little impressed with the botanical garden.[96] An English traveller who visited the university in 1775 states that "... you must not fail to see the Anatomy-chamber," but adds this is mainly because that is where "they preserve the money of Egypt; Pagan idols,—foreign dresses, birds from China, &c, &c." Again, not a word about anatomical preparations.[97]

So when Sandifort was appointed in 1771 as the successor of Albinus, he was facing a daunting task. Although hardly mentioned in histories of Leiden anatomy, Sandifort tried to lift both teaching and the collection up to modern standards while simultaneously attempting to do justice to the tradition of elegance. This shows in his preparations, as well as in his management of the collections and his ideas on teaching. In terms of teaching, Sandifort was

95 All calculations to current prices have been done using a historic calculator: http://www.iisg.nl/hpw/calculate.php, 22 July 2010.

96 Poole, *Journey from London*, vol. 2, 72–5.

97 Anonymous, *Holland*, 25.

eager to adopt the latest insights in teaching and to include pathology. Well into the eighteenth century, disease had been considered a misbalance of the Hippocratic humours, and the prescription of drugs was aimed primarily at rebalancing these humours so health would be restored. From the middle of the eighteenth century, the idea that disease might be located in a certain body part steadily gained ground, meaning that it became increasingly important for anatomists and medical practitioners to study diseased body parts. This shift obviously provided a great impetus for bedside teaching, post mortem research, and the creation of collections containing pathological anatomy.[98]

In 1761, at the age of eighty, the Italian anatomist Morgagni published his *De Sedibus, et Causis Morborum per Anatomen Indagatis*, a plea for the critical, experimental investigation of the root causes of deformations and pathologies, which would be reprinted well into the nineteenth century. The book soon spread across Europe and led Sandifort to declare a new programme for medical teaching. Hardly a decade after Morgagni's book was published, he held his inaugural address in Leiden. In *De Circumspecto Cadaverum Examine Optimo Practic Medicin Adminiculo*, Sandifort stated it is impossible to learn practical medicine without the cardinal knowledge of pathology, to be obtained by the study of the body through dissection.[99] And not only was dissection necessary for learning anatomy, both normal and pathological, it should also always be paired with the seeking of the laws of nature through the use of the senses and of reason:

> The physicians, servants and interpreters of nature, cannot reign nature if they do not know it, if they do not obey nature, even though they possess most knowledge. Observing, describing and obeying the laws of that same nature is what they exert themselves on, they spend their lives amidst experiments, and they strive with all their power to come to the utmost knowledge about diseased nature.[100]

98 Konert and Dietrich, "Giovanni Battista Morgagni."

99 Sandifort, *Oratio De Circumspecto Cadaverum*, 6: "Haec practicae medicinae Gubernatrix. Hac ad intimosabditosque organorum recessus paratur via, funestae mroborum exponuntur clades, tutior patet curandi methodus, & accuratior instituitur prognosis, ubi fatorum mandata rescindere non licet."

100 Ibid., 10: "Hi, medicum, naturae ministrum & interpretem, naturae non imperare posse, nisi hanc cognoscat, nisis naturae obtemperet, optime gnari. ipsius leges observare, notare, illis obsequy student, aetatem inter experimenta consumunt, & ad intimam morbosae naturae cognitionem pervenire, omni nituntur ope."

From this statement it becomes clear that the use of sensory perceptions and reason were also required to gain knowledge about pathology; further, there had to be more observations than that of a single case, and they had to be compared both to similar occurrences and to healthy anatomy. This was a shift away from the humoral understanding of disease and deformation towards a more localized one. Nonetheless, logic and aesthesis were still combined by Sandifort to gain anatomical knowledge, whether it was normal or pathological, ugly, wonderful, or beautiful. He was the first to give detailed descriptions of a congenital heart malformation that would later become known as the teratology of Fallot, and his Leiden successor Teunis Zaaijer (1837–1902) listed him as one of the great anatomists of the eighteenth century, among Cassebohm, Meckel and Sömmering.[101] Sandifort also had his findings concerning pathologies not only registered in minute descriptions but also in drawings and engravings, which led the French pathologist Jean Cruveilhier (1791–1874) to describe him as the "father of pathological iconography."[102]

Irrespective of the focus on pathology in his own research and teaching, Sandifort was a profound admirer and careful guardian of the Albinus collection. According to his son Gerard, his admiration for the subtle anatomical Albinus preparations was so great that he barely dared to practice it himself.[103] James Edward Smith, the founder and first president of the London Linnean Society, visited Leiden and met Sandifort during his tour of the continent in 1786. On that meeting, Smith wrote that "Professor Sandifort shewed me the Anatomical Theatre, and the preparations of Albinus; the latter can be seen in his presence only;" an indication of Sandifort's protectiveness of the Albinus collection.[104] That Sandifort tried to work in the same tradition of elegant, subtle anatomy as Albinus, also appears from how one of his Italian connections, the previously mentioned anatomist Paolo Mascagni, described him: as the successor of the immortal Albinus.[105] Mascagni did so in his 1784 work on the lymphatic system, which he tried to make visible through mercury injections, inspired amongst others, as we have seen, by Antony Nuck.[106]

Eduard Sandifort himself left no explicit clues as to his mercury preparations, but in the foreword to the first volume of the *Museum Anatomicum*

---

101 Zaaijer, "Het Gewigt," 21.

102 Luyendijk-Elshout, "The Elegant Anatomist," 319. For examples of Sandifort's pathological iconography, see illustrations 5.1 and 5.2.

103 Ibid., 312.

104 Smith, *A Sketch of a Tour*, 15.

105 Mascagni, *Prodrome D'un Ouvrage*, 132.

106 Mascagni, *Prodrome D'un Ouvrage*, "Introduction."

catalogue that he made upon the request of the board of curators of Leiden University in 1793, he states that Nuck in his time was said to be the most skillful in injecting the lymphatic vessels, and that he had traced all lymphatic vessels in the body—a bit of an exaggeration, but it had indeed been Nuck's aim to do so—with mercury mixed with lead or tin so it would harden. Sandifort concludes:

> Hence it is surprising that now that this part of Anatomy has finally reached a height in our time, after it had not been appropriately studied and even completely neglected for a while, our contemporaries refuse the honour of this [the discovery of hardening mercury injections, MH] to this man [Nuck], while he himself very rightfully believed he deserved it.[107]

The resentful tone is a bit strange, and raises the question which contemporaries was Sandifort talking about. It certainly cannot be Mascagni; as we just saw, he praised both Nuck and Sandifort himself extensively in his 1787 treatise on the lymphatic system. But this quote does show that Sandifort felt it unjust that the admired Nuck preparations had all but disappeared from the Leiden collections, and his admiration for Nuck explains why he probably decided to create mercury-injected preparations of the lymphatic vessels himself.

The materiality of the mercury-injected preparations in the Leiden collections also suggests that if they were indeed made by Sandifort, he made an effort to make them resemble Nuck's work as closely as possible. The first indication for this is the make-up of two preparations of mercury-injected, dried lymphatic vessels.[108] They are mounted on glass plates; originally those plates lay sunken into black-lacquered wooden boards that had to be destroyed in the late twentieth century for health and safety reasons. Because mercury was, and is, leaking from the preparations, the glass plates have been put in vessels and sealed of. But the original presentation is similar to the way Boerhaave described the display of Nuck's late seventeenth-century lymph preparations: injected with a mercury mass, dried, and mounted on lacquered wooden boards.[109] Given their resemblance to Boerhaave's discription of Nuck's preparations, it is not unlikely that these two preparations were indeed made by Nuck himself, and were at the source of Sandifort's admiration and his indignation about what he experienced as a lack of recognition for Nuck's work. The other mercury preparations support this hypothesis.

---

107 Sandifort, *Museum Anatomicum*, vol. I, xi.

108 Ak0005 and Ak0006. Also see illustration 3.0.

109 Boerhaave, *Praelectiones Academicae*, 576.

ILLUSTRATION 3.3 *LUMC Ag0022. Part of a liver, injected with mercury. Probably from the collection of Eduard Sandifort (1742–1814). © Anatomisch Museum LUMC 2012. Photographer: Arno Massee.*

The wet preparations can be divided into two groups: those of entire organs in which the lymphatics have been marked with mercury, and three of lymphatic vessels either in part of an organ or entirely separated.[110] The preparation of a part of the liver seems a bit of an oddity, because it is the only preparation of the lymphatic vessels in a part of an internal organ, and not in an entire organ or of separated lymph vessels like the others. Unfortunately, its original phial has also been lost, leaving us with very few clues as to why it may have been made. As Nuck first described in 1691, the liver is a particularly grateful organ for preparation, as its minute structure of vessels gives a spectacular visual effect when injected, so it is not unimaginable that this liver was one of Sandifort's first successful mercury-injecting projects.[111]

Two very similar mercury-injected preparations of lymph vessels from the belly show that although Sandifort was inspired by Nuck's work, he also

110 Al0007 heart of a child, Al0008 penis, Ag0022 part of a liver, and Ag0020, Ag0021 lymphatic vessels from the belly respectively.

111 Nuck, *Adenographia Et Uteri*, 144–5.

ILLUSTRATION 3.4 *LUMC Ag0021. Lymph vessels from the belly, injected with mercury. Probably from the collection of Eduard Sandifort (1742–1814).*

experimented with new techniques.[112] Unlike the previously discussed dried preparations of lymphatic vessels and the description of Nuck preparations, these preparations have not been dried, but preserved in oil of turpentine, which makes the tissue extra transparent, stressing the shimmer of the mercury injections even more. They are still in their original containers, enabling us to compare them to other preparations made by Sandifort. The finish of these two preparations resembles others from the Sandifort collection: the phials are the same size and shape, the lids and seals appear to be the same material as on others: turpentine as preservation fluid, cork stoppers sealed with a grayish wax. Sandifort therefore most likely made these preparations, still inspired by Nuck's idea of isolating and injecting the lymphatic system, but choosing a preservation method less fragile than drying and mounting them on wooden boards.

112 Ag0020 and Ag0021.

There are two more wet preparations of mercury-injected lymphatic vessels, not of separated lymphatic glands but of those in a child's heart and a penis.[113] Both these structures were also described by Nuck in his 1691 *Adenographia*, but he mentions nothing about injecting them—yet according to Ruysch the penis was one of the hardest structures of the body to inject successfully.[114] Al0008 seems to underscore this notion: the preparer has used too much pressure while injecting the glands, giving the entire glans penis a silver-plated look instead of only highlighting the miniscule vessels. Although the preservation method of these two preparations resembles that of the three Ag preparations, in a glass container with oil of turpentine, the phials are remarkably different in size and shape than the other Sandifort preparations. Moreover, these two have been sealed with what is most likely a slate plate and also a grey waxy lacquer, but with a very different shade of grey then the Ag preparations. Obviously this can simply mean that Sandifort made these preparations in a different period, but there is a small possibility that these preparations were Nuck's, and formed an inspiration for Sandifort.

Finally there are two more dry preparations injected with mercury, respectively of a piece of intestine and again a of child's heart, both injected with mercury and red wax.[115] Given the pristine condition these two preparations are in, it is unlikely that they belong to Nuck. Sandifort listed a wax-injected, dried preparation of intestines ascribed to Swammerdam in the Albinus collection in *Museum Anatomicum I*, but no mercury is mentioned in their description, which makes it more likely that Sandifort made these two preparations as well.

Assuming that five or even seven out of the nine mercury-injected preparations in the Leiden collections were made by Sandifort, why he made them yet subsequently did not list them as his work in the catalogues remains unclear. As we have seen, it is very well possible that this was somewhat of a private project for Sandifort, as it was not directly relevant for his main topics of research and teaching in Leiden—pathology and gross anatomy- but fitting for his admiration of the explorations and refined preparations of his illustrious predecessors, Nuck and Albinus. We have also seen that several aspects of aesthesis were important to Sandifort, such as gaining knowledge from sensory experience, materiality, hands-on practice, and reason, all directed towards working in a refined and elegant manner. It is in his mercury preparations that these factors come together most clearly, and it is not unlikely that he already started making them in the 1760s, when a renewed interest in lymphatic

113 Al0007 and Al0008. For Al0007, see Figure 3.0.

114 Nuck, *Adenographia Et Uteri*, 143, 147–8, Ruysch, *Effigies Penis Humani*.

115 Al0190 and a similar one without a catalogue number.

research arose; this may also have been why he kept them in his personal cabinet, showing them to interested colleagues and students every now and then.

The preparations clearly referred to Sandifort's illustrious predecessors: the mercury injections he chosewere in Nuck's style. By the 1770s, this choise had a dual function: on the one hand, mercury-injected preparations of the lymphatic glands of great interest once again as research on the lymphatic system and its function was far from completed, making. On the other, creating these preparations placed Sandifort firmly within the tradition that he called 'elegant anatomy.'[116] He went to quite some lengths to obtain his goals, even imitating the make-up of the lost Nuck preparations described by Boerhaave. Creating preparations in the stye of an illustrious predecessor was a tried manner to both celebate and criticize traditions and hypotheses, as we will also see in the next chapter. Sandifort used a notoriously difficult technique, and a material that was no longer immediately associated with obscure transmutational practices, to put himself on the same level as Albinus, while he was working on a current research topic. By the late 1780s, when Sandifort was busy compiling the first volume of the *Museum Anatomicum*, research on the lymphatic system had peaked again.

That there are so few of these preparations makes sense, as they were very difficult to create in the first place, and they remained very fragile when finished, easily damaged by changes in temperature. That preparations likely made by Sandifort were not listed in the *Museum Anatomicum* catalogues compiled by himself and his son is not surprising if the institutional collecting practices of the late eighteenth and early nineteenth century are taken into account. Until the first decades of the nineteenth century, professors of anatomy made and kept most of their preparations at home, and they were regarded as private property. Although a professor might occasionally bring preparations to the university for a class, or donate some preparations during his lifetime, when he died and his personal collection was not bequeathed to the university, it remained the property of the professor's descendants or was auctioned. Almost all preparations that were in the Leiden University anatomical collections until the 1820s were either donated by or purchased from anatomy professors upon their deaths. Only in the nineteenth century collection ownership gradually shifted from private to institutional.[117]

When Sandifort started compiling the *Museum Anatomicum*, the University curators wanted catalogues listing the institutionally owned collections: the historical collections that were deemed relevant for teaching, or due to their

116 Luyendijk-Elshout, "The Elegant Anatomist."

117 Huistra, "Preparations On The Move," 20.

historical value, and of the new collections that showed how up-to-date Leiden University's department of medicine was.[118] No wonder that Sandifort's small, probably privately owned, eclectic, and slightly anachronistic collection of mercury-injected preparations of the lymphatic system, however pretty and refined, was not mentioned in those catalogues. By the time his son and successor, Gerard Sandifort, compiled volumes 3 and 4 of the *Museum Anatomicum* over thirty years later, interest in historical preparations had withered. The 1827 third volume of the *Museum Anatomicum* primarily served to describe the newly acquired collections of Bonn and Brugmans. Furthermore, in the preface to the fourth volume of the *Museum Anatomicum*, Gerard Sandifort states that in that work he has tried to list the most important specimens of pathological anatomy from the collection, with illustrations, claiming it is pathology that is hardest to learn for the anatomist. After all, when dissecting, one can expect to encounter mostly normal anatomy, but with pathology, especially that of rare diseases, one cannot expect to learn through dissection only.[119] Although Gerard might have appreciated the few specimens of elegant anatomy his father made, and although they might have been housed in the anatomical theatre by that time, they were not part of the 'core business' of Leiden anatomy teaching around 1835.

Another reason no attention was paid to these preparations in the first decades of the nineteenth century may have been that by then mercury was becoming less popular as an anatomical injection mass. The penetrative quality turned out to be a disadvantage in certain cases: mercury frequenly broke through the microscopic barriers between the lymphatic glands and the veins, leading to incorrect ideas about their relation.[120] Combined with the fact that mercury has to be mixed correctly with hardening agents in order to prevent damage to the finished preparation, due to temperature changes that cause the mercury to expand and retract, it would become increasingly unpopular as an injection mass in the course of the nineteenth century. These combined factors made the mercury-injected preparations created by Sandifort senior irrelevant for the listings in the last two volumes of the *Museum Anatomicum*. It is not likely that Sandifort junior omitted the preparations from the catalogue because he thought they referred to pre-eighteenth-century chymical symbolism. Mercury's meaning changed in the course of the eighteenth century, losing, or at least loosening, its ties with many of its older, transmutational connotations. But mercury was still an accepted injection mass for anatomical

118 A.C. 228, 26 April 1798. Transcription in Witkam, "Over De Anatomieplaats," appendix 38.

119 G. Sandifort, *Museum Anatomicum*, vol. IV, 1–2.

120 Knox, *The Anatomist's Instructor*, 122, Cole, "The History of Anatomical Injections," 341–2.

preparations because of its material and visual qualities, as is also shown from the previously mentioned ongoing listings of mercury preparations in eighteenth-century anatomical collections.

Besides, even though the downsides of mercury injections became more important in the nineteenth century, being able to inject lymphatic vessels with mercury remained a proof of skill. From the late eighteenth century onwards, the powerful group of practitioners who relied upon experimental skill and craftsmanship, like Eduard Sandifort, was steadily growing and stimulating the incorporation of hands-on experience in higher education, supported by administrators.[121] But mercury as an anatomical injection mass remained an acquired and fairly exclusive skill: in his 1860 preparation handbook, Hyrtl remarks that it is "no job for students," who "except in class have never seen lymphatic vessels, let alone that they wish profoundly to try and inject them." According to Hyrtl, there are few anatomists who occupy themselves with this 'specialism.'[122] Nonetheless, Dutch anatomists would continue to make elegant mercury-injected preparations well into the nineteenth century: Zaaijer described the mercury-injected preparations of his predecessor Halbertsma as 'refined' in his inaugural lecture of 1866, and former Leiden student Vrolik kept the tradition of injecting preparations with coloured wax and mercury alive until his death in 1880.[123] In this light, it is not surprising that, eventually, some of the elegant preparations made by Eduard Sandifort reappeared in an 1866 collection description by G.C.B. Suringar.[124]

Thus an analysis of the use of mercury injections in anatomical preparations in the late seventeenth century and the second half of the eighteenth century shows the material aspects of the aesthesis complex of properties, experiences, and activities. Eduard Sandifort's reasons for creating mercury-injected preparations of the lymphatic system in the late eighteenth century stemmed from a deep admiration for the technical, practical injecting and preparation skills of his predecessors, and a wish to both be placed in the tradition of aesthesis in Leiden and to be a modern anatomist. The characteristics of aesthesis help us understand the eighteenth-century practice of injecting lymph glands with mercury. First, the silvery shimmer of mercury beautifies preparations of otherwise rather unsightly parts of body fabric. Following on from that, creating mercury-injected preparations was a hard-won skill, the kind of tacit bodily

121 Klein and Spary, "Introduction: Why Materials?," 7.

122 Hyrtl, *Handboek Der Praktische Ontleedkunst*, 655.

123 Dr. Laurens de Rooy, curator of the Amsterdam Museum Vrolik, pointed this out to me.

124 Luyendijk-Elshout, "The Elegant Anatomist," 318.

knowledge that can only be acquired through endless hands-on experience. That also suggests the last reason why the use of mercury fitted into the epistemic culture of aesthesis: these preparations are so refined and elegant, yet at the same time so undeniably a product of sensory experience and experiment combined with reason, that their beauty is not simply in their shimmer, but in their unlikely material perfection, their deeper historical and symbolic meanings, and in the new knowledge they represent too.

Moreover, this chapter has made it clear that Eduard Sandifort's mercury preparations can only be understood through the history of the practice of using mercury as an injection mass for anatomical preparations; a history strongly intertwined with mercury's material meaning in transmutational alchemy, the latter's erratic removal from chemistry, and medicinal uses. Mercury was the injection mass of choice for the mysterious, difficult to locate, and supposedly cold lymphatic system. It was a fixing principle as well as a resistant, hard to handle material. Rooted in mid-seventeenth-century experiments and originally sought for its symbolic value, penetrating, resurrecting, and fixing qualities, as well as its attractive shimmer, mercury-injecting not only befitted aesthesis, but also referred to a host of historical and contemporary uses and meanings in medicine, mythology, and chemistry. Another reason for the popularity of mercury preparations, despite the difficulty of making them, was that they made it possible to permanently visualize a normally invisible and, until then, largely unknown bodily structure, the lymphatic system, which would have been difficult to make visible or accurately describe otherwise.

Injecting and discovering new structures could make the anatomist immortal, as Reinier de Graaf pointed out. Yet the immediate Leiden successors of Nuck, Ruysch and Albinus were occupied with other anatomical issues than that of the lymphatic system and thus created preparations with materials other than quicksilver. For their life-like preparations, coloured wax was a more appropriate injection mass, and they also needed other materials, like lace-rimmed sleeves, as we will see in the next chapter. Yet mercury-injected preparations of the lymphatic system were made, collected, and admired in many places throughout the eighteenth century. Thus it is no surprise that Eduard Sandifort picked up where his famous predecessor left off, and successfully tried to include himself in the tradition of elegant anatomy by creating and improving mercury-injected preparations of the, still largely mysterious, lymphatic system. After all, the miniscule lines of injected mercury in a preparation, now devoid of transmutational symbolism, remained a proof of the elegance of the anatomist who created it; of his refined taste, sense of beauty, and hard-to-match skill in making such a detailed and fragile object.

It turns out that the fact that Eduard Sandifort's mercury preparations were not listed in the *Museum Anatomicum* catalogues of the Leiden university anatomical collection published by him and his son had practical reasons. When the first two volumes were published the preparations probably did not yet even exist, and the second two volumes, published after Sandifort senior died, simply had a different aim: they were meant to document the recently-acquired anatomical and pathological collections available for teaching at Leiden University, and not to give an overview of everything present in the Leiden anatomical collections. Because Sandifort's collections long remained largely undocumented private property, they were initially thought to be Nuck's. But they were in fact the latest products of aesthesis in early nineteenth-century Leiden, and creating mercury-injected preparations remained a proof of the skill and elegance of the anatomist well into the nineteenth century.

CHAPTER 4

Hands, Lace and Plants: Meaningful Embellishments

Two infant hands in jars, lifelike in lace sleeves, hands rising upwards in their phials, one of them holding a flimsy red piece of tissue on a string, the other lifts a vulva tied to a lacy ribbon. Another phial holds a fragment of a plant, a sole frumpy flower on one of its branches. On the bottom is a ghostly collection of white flakes.[1] The first two preparations are on permanent display in the Leiden Museum Boerhaave and the Leiden University Anatomical Museum respectively. The last preparation, the phial with the white flakes, sits in a storage cabinet in the cellars of the Leiden University Medical Center. A first superficial description of the materiality and location of these three particular preparations raises more questions than it explains. Why do those hands look so life-like and pinkish, why are they holding bits of tissue and why are they wearing the lace sleeves? What is the thing on the string, and why would one make a preparation of a child's hand holding a vulva? What are the white flakes on the bottom of the phial, and why is there a flowered branch in it? Do these preparations have something to do with one another; do they even fit into some kind of wider tradition? How were they looked at when they were first made and what happened to them as time went by?

One of these preparations (the hand holding the vulva) is ascribed to the Amsterdam anatomist Frederik Ruysch (1638–1731), the other two to his student and Leiden anatomy professor Bernard Siegfried Albinus (1697–1770). Relatively much has been written about Ruysch and his collections: A.M. Luyendijk-Elshout in 1970 was the first to relate the decorations of the Ruysch preparations to contemporary vanitas ideas about life, death and the human body.[2] Art historian Julie Hansen rather anachronistically identified Ruysch primarily as an artist in a 1996 article.[3] Luuk Kooijmans published an extensive biography of Ruysch in 2004 that appeared in translation in 2009, Gijsbert de Roemer explored possible theological connections in an article in the same year, and Josien Driessen van het Reve explored the faith of the Ruysch collections after they were incorporated in the St. Petersburg Kunstkamera in 1717

---

1 LUMC catalogue, numbers Ab0015, Al0009, Ab0001.

2 Luyendijk-Elshout, "Death Enlightened."

3 Hansen, "Resurrecting Death."

 | DOI 10.1163/9789004262775_005

in an exhibition catalogue and article in 2004 and 2005 respectively.[4] Dániel Margócsy discussed the advertisement of anatomical collections and skills by anatomists like Ruysch in 2008, and Rina Knoeff recently published an article on Ruysch' display strategies.[5] Albinus also had his fair share of attention, mainly in the work of Luyendijk-Elshout and Punt between 1950 and 1985, yet no attempts have so far been made to understand the resemblances and differences between their preparations, nor has an object-driven analysis of the Albinus preparations been attempted before.

This chapter shows that the Ruysch and Albinus preparations are actually closely connected, but also distinctly different, and that a close reading of their materiality reveals even more about their make-up and meaning. First, the use of coloured wax injections in anatomical preparations in this period will be explored, particularly the use of red pigments. Subsequently, the arms and the tissue they are 'holding' are discussed, as well as the possible uses and meanings of the lace-rimmed sleeves with which they are decorated. This analysis will provide an initial understanding of the influence of aesthesis on the creation of preparation in which hands are combined with other body parts and textiles. After that, the mysterious preparation of flakes and a branch will demonstrate that deciphering the aesthesis of this anatomy is sometimes far from straightforward. Finally, the reception history of these preparations will be discussed briefly in order to gain some more insight to the problems and opportunities these preparations presented and present.

Materies rubra, materies coerula: Coloured Wax Injections

The two arms preserved in the preparations by Ruysch and Albinus look surprisingly healthy and lifelike in their phials, with a pinkish glow—a far cry from most other wet preparations of limbs, which have become ghostly white under the influence of their preservation liquid. The secret to their glow is in arterial and venal injections with a hardening wax mass coloured with red pigment. The choice of red pigments to achieve a lifelike effect and to visualise veins and arteries may seem obvious because of the apparently timeless connotations with red blood and healthy blushes, but there is more to this technique than meets the eye. As Domenico Bertolini Meli points out, colour is one of the most immediate sensory experiences yet also one of the most complex philosophical

4 Kooijmans, *De Doodskunstenaar, Death Defied,* De Roemer, "From Vanitas to Veneration," Driessen-Van het Reve "Geloof Alleen Je Eigen Ogen," "Frederik Ruysch in Rusland."

5 Margócsy, "Advertising Cadavers," Knoeff, "Sex in Public."

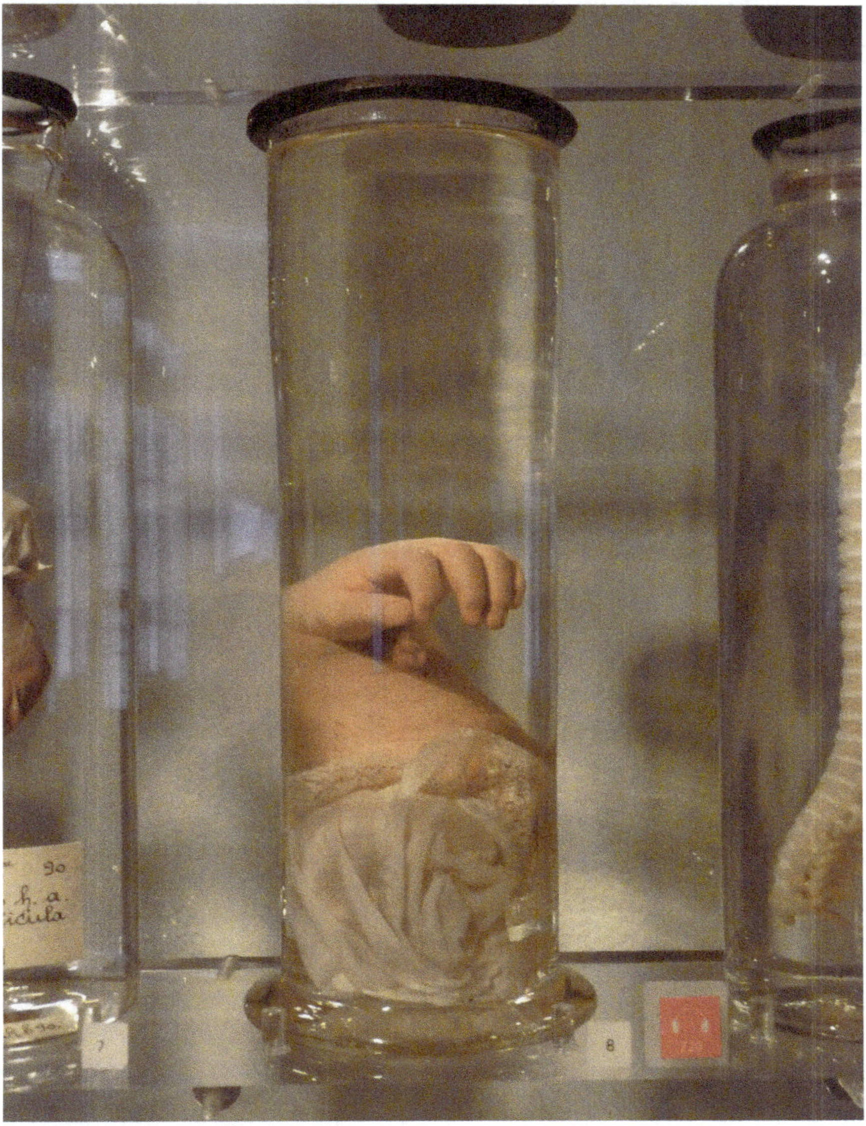

ILLUSTRATION 4.0 *LUMC Ab0015. B.S. Albinus (1692–1770). Child's arm with lace-rimmed sleeve, holding a choroid membrane. Injected with red wax. Photograph: author.*

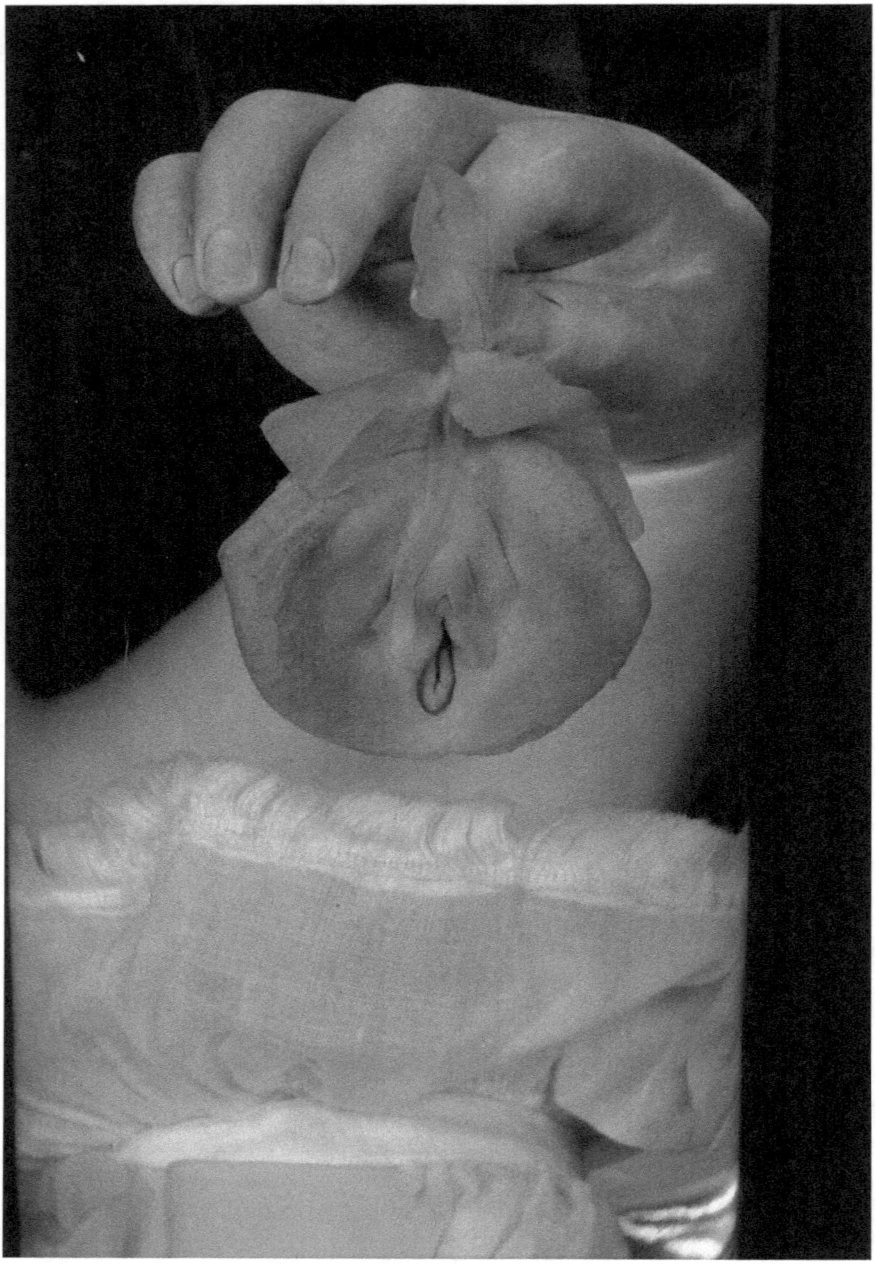

ILLUSTRATION 4.1 *LUMC Al0009. Frederik Ruysch (1638–1731). Child's hand holding a vulva. Injected with red wax. © Anatomisch Museum LUMC 2012.*

and physiological problems in sense perception, so redness should not be taken as an inherent and universally understood quality of blood.[6] Blood had diverse and changing cultural, religious, medical, and physiological meanings in the early modern period.[7] For example, blood was not a research object in anatomy until William Harvey (1578–1657) made the discovery of the circulation of the blood in 1628.[8] On top of this, the hermeneutics of colour are very complex. As we have seen previously, red was associated with hot, dry, active sulphur; the opposite of cool, wet mercury well into the seventeenth century.[9] Moreover, the intrinsic redness of blood was not discovered until the late seventeenth century, and certain kinds of red pigments turned out to be more suitable for resembling blood in anatomical preparations than others.

Jan Swammerdam developed injections of anatomical preparations with hardening coloured wax in the 1660s.[10] This technique enabled anatomists to create lasting preparations that showed the structures they had discovered through their injection experiments. In the initial injection experiments, all kinds of coloured fluids were used to visualize bodily structures: black, yellow, green, and even milk was used because of its clear white colour. This led Reinier de Graaf to the conclusion that all the parts of the body were coloured more or less red by the blood. Previously, it had been thought that it was the liver that changed white chyle into red blood, a vivid analogy with the process of rubedo, the rubification or reddening of the white matter of the Philosopher's Stone at the final stage in the opus alchymicum.[11] Descartes for example, in his 1662 treatise on man in neo-Galenic style argued that white chyle is transformed into red blood in the liver like the white juice of black grapes is turned into red wine.[12] Throughout the seventeenth century, a vivid discourse on the colour of blood emerged throughout Europe. Learned men such as Malpighi, Fracassati, Borelli, Boyle, Hooke and Lower studied the problem.[13] But Reinier de Graaf concluded that a washed-out liver injected with milk actually turned

---

6 Bertoloni Meli, "The Color of Blood," 118.

7 See i.e. Santing and Touber (eds), *Blood-Symbol-Liquid*.

8 Cunningham, "The Principality of Blood," 196–7.

9 See chapter 3. Also Abraham, *A Dictionary*, 192–3 and Bucklow, *The Alchemy of Paint*, 86.

10 For a more extensive discussion of the development of techniques to create lasting anatomical preparations see chapter 3.

11 Abraham, *A Dictionary*, 174–5.

12 Bertoloni Meli, "The Color of Blood," 120–1. The *Traitté de l'homme* was actually written between 1630 and 1633, but only published after Descartes' death in 1662, likely because he feared papal prosecution.

13 Bertoloni Meli, "The Color of Blood."

white, and concluded that the redness of blood must be an intrinsic quality.[14] De Graaf's discovery that blood is actually red of itself and does not derive its colour from the liver gave blood a new, autonomous, life-giving status.

This distinction of the red blood as a life-giving force made red the colour of choice for the hardening injection mass for anatomical preparations in the late seventeenth and early eighteenth centuries, especially for preparations like those made by Ruysch and Albinus; preparations that simultaneously clarified anatomical structures and served as life-like preservations of entire children, extremities, and heads. Other colours were only used when two different structures otherwise hard to distinguish needed to be made visible in a preparation, and there were no agreed conventions about which colour should be used for which purpose. For example Blankaart, in his 1695 anatomical handbook that would be reprinted well into the eighteenth century, is very brief on coloured wax injections: they should be made out of white wax, or sheep's tallow and oil of turpentine, coloured with 'any pigment you like,' like filtered vermillion (red) or rust of copper (green), to distinguish between the arteries and the veins. The anatomist should make sure the mass is fluid and will flow to the smallest branches.[15] That is all. No quantities, no temperatures.

Yet the obvious problem with the coloured waxy substances used for injecting anatomical preparations is that they have to solidify at room temperature in order to obtain a lasting preparation, but keeping both the preparation and the injection mass warm enough for the mass to remain fluid during injection is very difficult. Otherwise, the injection mass congeals before it reaches the smallest branches of the arteries or veins. Adding oil of turpentine can to some extent solve this, but if too much of it is added, the mass will not solidify at room temperature any more. Injecting a preparation with coloured wax in such a way that the wax reaches even the smallest veins before it hardens, thus visualizing otherwise indiscernible structures or, in the case of preparations of an entire limb, giving it a life-like, pinkish glow, is extremely difficult. Finding the right mixture of ingredients and the right temperature and pressure for injecting can only be done through endless trial and error. It is an intensely

14 De Graaf, *Korte Beschryving*, 665–6.

15 Blankaart, *Anatomia Reformata*, 1695, 758: "Cerae praeparatio XXIX. Cera hoc modo praeparatur. Recipe ceram albam, cui sevum ovillum junge: hisce simul liquefactis spiritum terebintinae adde. Pigmentum adde quod lubet, ex minio, aerugine aliove colore pro re nara (distinctionis gratia inter arterias & venas) suspenditur in syndone tenui; alias materia liquefacta colari debet, quando pigmenta cerae liquefactae colari debet, quando pigmenta cerae liquefactae admiscentur. Ne ob particulas crassas cananicoli obstruantur, atque injectio usque ad minima vascula inhibeatur."

physical and sensory experience: after a couple of hours of trying to make a wax-injected preparation, your back and arms hurt from holding the specimen and the syringe, your hands are cramped and possibly burnt from the hot wax, your eyes are strained from searching for the minuscule veins you want to fill, your sense of smell is overwhelmed by the stink of the dead body and the wax, and you feel light-headed from the fumes of heated tallow and turpentine.

Nonetheless, if an anatomist was successful in creating lasting wax-injected preparations, the results could be stunning. Ruysch was the first to successfully inject many anatomical preparations with coloured wax, and his preparations brought him financial wealth and academic as well as social esteem. The preparations objectified the dead body; making a preparation was making a tradable commodity, an object that could serve as a research result, a teaching aid, a moral agent, and a curiosity at the same time. Initially Ruysch had appeared unwilling to part with his recipes for preservation and injection fluids, but when he sold his cabinet to Tsar Peter the Great in 1717, he included the recipes in the sale. The knowledge that Ruysch's recipes were no longer in his possession ignited an international quest for them; an indication of how highly valued they were. However, Ruysch no longer seemed to worry about it, as he had realized that it was not so much the recipes as his skills that were the key to his unique preparations.[16] He concluded that creating injected and lasting preparations was "a secret that only I know. Because whoever I asked, and whatever I found out, there is no one else who really can do it."[17] This illustrates that making anatomical preparations rested on tacit, bodily knowledge from the early days, and that anatomists were well aware of this.

The tacit nature of this knowledge also shows in the work of the man who made the other preparation, that of the hand holding eyelids and a choroid membrane on a string—Bernard Siegfried Albinus (1697–1770). He was a student of Ruysch, and professor of anatomy and surgery at Leiden University from 1719–1770. The young Albinus was not at all the outgoing entrepreneur Ruysch was, but he was hungry for knowledge about the smallest structural element of the body and the source of life, and therefore followed Ruysch's example. He too vigorously explored the possibilities of the fairly new technique of vascular injection with hardening coloured substances to visualize the structures of the human body. He hoped that studying those structures

16 On the sale of Ruysch's cabinet and the subsequent hunt for his recipes, see Kooijmans, *De Doodskunstenaar*, 318–23.

17 Ruysch quoted in Kooijmans, *De Doodskunstenaar*, 323: "... een geheim dat slechts ik alleen ken. Want wie ik er ook naar heb gevraagd en wat ik er ook over te weten ben gekomen, er is niemand die het echt kan."

through injecting would help him in discovering the smallest structural elements of the body, as well as tell him more about the functions of the bodily fluids and structures like the nerves.[18]

In his lectures, Albinus mentioned three different substances for injecting anatomical preparations, but he never specified their composition—once again probably not because he regarded this as a valuable secret, but because he felt that it was not so much exact recipes as skills that were the key to making wax-injected anatomical preparations. Albinus mentioned red wax—materies rubra, materies ceracea or liquor rubor- mainly used for the injection of the arteries, and blue wax or 'materies coerulea' that he used for injections of veins.[19] Albinus had apparently found a mixture that overcame the difficulty of premature hardening of the wax, as he managed to inject even the smallest vascular structures through the bigger ones. Punt suggested it may have been the addition of some kind of resin, as that can still be smelled when the bottles are opened up for maintenance.[20]

Albinus never specified the red pigments he used, but according to Punt they must have been cinnabar and red lead.[21] Cinnabar is a quite common, naturally occurring pigment that can also be manufactured by reacting mercury with molten sulphur, a spectacular-looking but dangerous process that emits poisonous fumes.[22] Until the late seventeenth century, the words cinnabar and vermillion were used interchangeably, although medieval and early modern artists preferred manufactured vermilion over naturally-occurring cinnabar, as it was considered to be purer, a step higher on the ladder to perfection than cinnabar.[23]

But if vermilion was considered to be purer than cinnabar, it seems strange that Albinus, who was so fond of perfection, may have chosen cinnabar instead of vermilion. We will probably never know whether he indeed used only the natural pigment cinnabar and not the manufactured substance for economical and practical reasons. As for red lead: this pigment does not occur frequently in nature, but was already produced by the Romans through calcination of white lead, and was quite commonly used as a pigment in the illumination

18 Punt, *B.S. Albinus*, 75.
19 Ibid., 76.
20 Ibid., 76–8.
21 Ibid., 78. Hyrtl, *Handboek Der Praktische Ontleedkunst*, 532, also concluded that Albinus never specified his injection mass.
22 Bucklow, *The Alchemy of Paint*, 86.
23 Ibid., pp. 86–9, 224–46.

of manuscripts in the Middle Ages and Early Modern period.[24] Both cinnabar and red lead were known to be toxic by the early eighteenth century.[25]

Anatomists may have preferred red lead as a red colorant to cinnabar for both alchemical-philosophical and practical reasons. In alchemy, lead was known as a relatively heavy metal, just like mercury. The heavier the metal, the more perfect it was, with gold as the pinnacle of perfection.[26] One practical benefit of red lead as an injection mass may have been that its heaviness was helpful in the penetration of small veins. The other reason for choosing red lead over cinnabar may have been that unlike the latter, red lead is colourfast, does not fade in daylight, hardly dissolves in water or alcohol, and does not turn into acrid yellow fumes when heated like manufactured vermillion does—a benefit for preparations injected with heated wax substances and preserved in some kind of water-alcohol solution.[27]

Most likely, Albinus' choice of red pigments was prompted by a combination of their colour and their material qualities: the colour red was inextricably connected to qualities of hotness and wetness, as well as to blood and its possible life-giving forces. Both cinnabar and red lead were pigments that had long been known and used in both chemistry and art, they could be fairly easily obtained and they had particular qualities, such as limited solubility in the case of red lead, that made them suitable for the use in anatomical preparations that would be preserved in liquid. Yet given the durability of red lead, it is not unlikely that Albinus preferred this to cinnabar.

What is That Thing on a String? Hands Holding Things

Now that we know more about the inside of these preparations, we can move on to their outer material qualities. The first analysis is of the curious preparation of a child's arm holding a piece of tissue on a string like it is some sort of pendant.[28] Whether you think this preparation beautiful or horrific, or maybe both, there is more to it than meets the eye. Considering the preparation still looks life-like after almost three centuries, the anatomist did an amazing job here. Although it is difficult to tell from the picture, the close observer can see this left arm of a newborn with a lace-decorated sleeve has been stripped of

24 West Fitzhugh, "Red Lead and Minimum," 113.
25 Swiderski, *Quicksilver*, 43, 165.
26 Abraham, *A Dictionary*, 86–8.
27 Bucklow, *The Alchemy of Paint*, 245, Gettens et al., "Vermilion and Cinnabar," 59.
28 LUMC catalogue, number Ab0015. Sandifort, *Museum Anatomicum*, vol. 1, 29.

its skin and nails. As for the thing on the string, it is only from very close and with considerable anatomical knowledge that the observer would be able to tell that this is a piece of the choroid membrane with the iris, of which the miniscule veins have been injected with red colorant as well. But what neither the old nor the new catalogue descriptions of this preparation answer are the pressing questions the physical object raises. Why would an eighteenth-century anatomist create such a particular preparation of a skinned hand and an eye? Surely not merely to demonstrate his skills with the scalpel and syringe, or to create a visually arresting piece.

Albinus is probably best known for his much-copied and discussed anatomical atlas, the *Tabulae*. But as Albinus believed that man was God's supreme creature in which the *vis vitalis* or life spirit was the most forceful and perfect of all living creatures, he was striving to show this elegance and perfection in all his work. His quest can be seen in his refined injected preparations, in his selections of 'perfect' symmetrical specimens for the drawings in his anatomical atlas, and allegedly he was capable of performing dissections so precisely that not so much as a hair was misplaced, and in such a way they equalled the etchings in Eustachio's tables in their precision.[29] However, the focus is now exclusively on Albinus' anatomical preparations. Trained by the illustrious professor Herman Boerhaave and the famous Frederik Ruysch, Albinus stood in a tradition which combined anatomical observation and practical experiment with reason to develop physiological knowledge. This tradition started in the sixteenth century when Andreas Vesalius (1514–1564) was the first professor of anatomy to take up the scalpel and perform dissections himself, instead of leaving the dirty work to a surgeon prosector. This led to the testing and adjustment of the previously dominant anatomical knowledge passed on by the Greek Galen. Anatomy was no longer primarily philosophy, but an art, which combined with physiological philosophy, gave new insights to the structure and functions of the human body.[30] From then on, the anatomist's hands were his most important instruments, something also made clear through the iconography of the title page of Vesalius' 1543 anatomical atlas, *De humani corporis fabrica*. Yet the simultaneous quest for beauty and perfection in anatomy was at its height in the work of Albinus and his contemporaries.

29 Punt, *B.S. Albinus*, 220–2. Von Haller, "Albrecht Hallers Tagebücher," 93, notes on Albinus: "Er zergliedert schon izt mit solcher Richtigkeit, daß kein Häärgen versehrt wird, zeigt der Nerven Gestelle so gut, als es in Eustachio abgestochen . . .".

30 For the differences and interconnections between anatomy and physiology in the eighteenth century see Cunningham, "The Pen and the Sword. Part One. Part Two."

ILLUSTRATION 4.2 *Portrait of Vesalius in his 1543 anatomical atlas,* De humani corporis fabrica.

So what were Albinus' reasons to make this preparation of a hand with a lace-rimmed sleeve holding a piece of an eye, apart from placing himself within a tradition? Steinke has recently noted that Albinus' importance in the history of physiology is difficult to assess, as he never published any substantial work on physiology—his reputation as a physiologist rests on his lectures on 'human nature,' which he delivered as an anatomy professor.[31] This might be explained by the fact that for Albinus, anatomical observations and sensory experiences were the basis for physiological knowledge. When trying to answer physiological questions he primarily tried to do so by starting with anatomical analysis, that is, he studied the structures of the body in situ, hands-on, through

31 Steinke, *Irritating Experiments*, 36.

dissection and preparation. Only after that he would move on to the theoretical disciplines of physiological experiment and experimental argumentation, which meant philosophical reasoning about structures and functions. This approach of old anatomy combined with old physiology, which has been vividly described by Cunningham, resulted in a world famous collection of over three hundred anatomical preparations, about 180 of which survive today.[32]

Albinus' fascination with observation and sensory experience is reflected in his collection as well. Relatively few preparations were of the nerves (only 3 according to the 1793 catalogue), but those of the sensory organs, especially of eyes and skin, were numerous: about 36% of the total number of preparations. The preparations of the sensory organs and of the nerves should be considered as one collection, as to Albinus they were both parts of the same classical physiological category: the animal functions.[33] As appears from lecture notes taken by his students, Albinus started his lecture series on the animal functions with the physiology of the nervous system: first the brain and the nerves, then subsequently he discussed the sensory organs—the epidermis as well as the nails and hair, the tongue, nose, eyes and ears.[34] The collection of sensory preparations is particularly interesting to study as more than half of it has survived, and because it was clearly important to Albinus in developing and teaching his physiological ideas.

As for preparation Ab0015, its significance is fourfold. First, it is an investigation of two organs of sensory perception as discussed in Albinus' lectures: the skin (touch) and the eye (sight). Second, it is an anatomical masterpiece, a demonstration of Albinus' skill and elegance. It is refined, lifelike, and extremely difficult to prepare. Albinus himself could get very excited about a delicate preparation like this one, as appears from a note in a letter to his friend Robert Nesbitt, the English anatomist, in 1721. Jotted between a remark on shipped books and his farewells the young Albinus wrote:

> I should not forget to tell you that I have [...] an intact epidermis of the hand of a four-month-old foetus, which is very thin and very white and

32 Cunningham, "The Pen and the Sword. Part One. Part Two."

33 Albinus divided his physiological lectures after 1740 according to the classical *humores*: there were the vital functions (composition of the blood, structure and function of the heart, arteries and glands), the animal functions (nervous system, sensory organs, muscles), and the natural functions (respiration, digestion, nutrition, conception).

34 Punt, *B.S. Albinus*, 135. Box 1742, vols II & III, 229–451.

> resembles a small glove, it has not even the size of half of the nail of my little finger.[35]

His pride and excitement regarding the refinement of this object are patently obvious. Third, this preparation is an allegory. As noted previously, the hand was considered the anatomist's most important instrument since Vesalius. But the hand would be lost without the eye. Sight was traditionally considered to be the highest and most reliable sense, and it was hugely important in gaining anatomical and physiological knowledge. This importance also shows from the amount of attention paid to it: almost one hundred of the two hundred and sixty pages on the animal functions in Box's lecture notes of Albinus' classes are devoted to sight. Therefore, it is safe to assume that preparation Ab0015 was not only an investigation of two sensory organs, namely the skin, with nails attached to it, and sight, but that it was also intended as a demonstration of how to elegantly dissect the epidermis and nails, and further as an allegorical reference to anatomy's two most important senses for gaining knowledge: touch and sight. This combination of anatomical investigation and allegorical reference was not something Albinus had come up with himself. One of his teachers, Frederik Ruysch, was the actual inventor and master of the genre—and this leads us to the fourth dimension of this preparation.

Ruysch (1638–1731) started his career as an apothecary, studied anatomy in Leiden under Sylvius, and held numerous posts during his long life, such as praelector to the Amsterdam surgeon's guild, supervisor of the city's midwives, and forensic court physician. Although he never lectured in Leiden, he frequently visited the city where he had famous friends, such as Boerhaave, and adversaries—like Rau. Leiden students like Albinus also travelled to Amsterdam to see Ruysch's private collections and to attend his lectures and anatomical demonstrations. Ruysch's enormous anatomical cabinet in his Amsterdam home, which at its heyday numbered about 2,000 preparations, contained many objects which not only testified to his anatomical inquiries and findings, but which were also allegorical masterpieces. The other preparation of a hand discussed here, the one holding the bow-tied vulva, is from his collection. It is one of the few preparations by Ruysch found in Leiden today, as his entire cabinet was purchased and shipped to Russia by tsar Peter the Great in 1717. These preparations are now housed in the St. Petersburg Kunstkamera,

35 Transcript of letter from Albinus to Nesbitt, 28 October 1721, in Le Fanu, "More Letters": "Il ne faut pas que j'oubli[e] de vous dire que j'a[i e]piderme entiere de la m[ai]n d'un foetus de 4 mois, qui tres mince & tres blanche ressemble a un petit gand, qui n'a pas la grandeur de la moitié de l'ongle de mon petit doigt."

and many of them can be studied online today.[36] Thanks to Ruysch's extraordinary skills, his preparations are still there for the world to see after three hundred years.[37] What then, can we learn about Ruysch's intentions in both anatomical and allegorical respects, from studying this preparation?[38]

There are many more preparations of hands in Ruysch' St. Petersburg collection and described in his *Thesaurus* that are not alone in their phials: there are children's hands holding pieces of lung, of testis, fragments of dura maters and placentas, a breast, a kidney, a foetus, the 'skeletons' of a pear and an oak leaf, a calf spleen, a small fish, a passion flower and fruit, and one is even accompanied by a scorpion.[39] It seems a collection of randomly combined objects in which the children's hands, judging by the descriptions Ruysch himself gave them, serve either as specimens of the structure of the skin or simply as a kind of display tool to present, commodify, and objectify all kinds of bodily structures and exotica. The hands, looking like they are still alive by holding out other specimens as if they are artefacts, cause this commodification and objectification: they present the specimens as if they are not body parts or botanical or zoological things, but independent objects, suitable for trade and exchange. Margócsy has also described this process in a 2008 article: Amsterdam anatomists like Ruysch 'advertised' their anatomical preparations in print.[40] The strategy of commodification would continue to be used by generations of future anatomists.[41] But why make a preparation of a hand holding a vulva? Anatomical preparations of genitals are somehow doubly disgusting: not only are they representing the incomplete, mutilated, dead body, they are also reminding us of 'disgusting' bodily functions, such as defecating and procreation, functions that make the body part of the cycle of life and death, of eternal recurrence.[42] To understand why Ruysch made this preparation and how it was perceived by his contemporaries, I will explore some other preparations of female genitals in his collection descriptions, and discuss a controversy in which he was involved.

---

36 http://www.kunstkamera.ru/kunst-catalogue/index.seam?c=RUYSH&page=1.

37 The Ruysch collections in the Kunstkamera have been lovingly restored in the past decade, but were neglected for centuries after Tsar Peter the Great died.

38 LUMC catalogue, number Al0009.

39 See http://www.kunstkamera.ru/kunst-catalogue/index.seam?c=RUYSH for the digital collections.

40 Margócsy, "Advertising Cadavers."

41 See chapter 6 particularly for this.

42 Miller, *The Anatomy of Disgust*, 40–1, Korsmeyer, *Savoring Disgust*, 35.

Why would Ruysch make preparations of female genitalia anyway? The most straightforward answer is: because all his contemporaries who got the chance also did so. The organs of generation were still largely a mystery in Ruysch's time, so anatomists were keen on dissecting genitals and reproductive systems, including female genitals and the uteri of pregnant women, in order to find out more about the differences between men and women and how children were conceived and developed in the womb.[43] Through such dissections, Ruysch discovered a muscle in the womb that he held responsible for the extraction of the placenta after birth.[44] His *Thesaurus* mentions numerous other preparations of the female genitals, apart from the Leiden preparation discussed here. When one starts reading the descriptions Ruysch gave of these preparations, something stands out immediately.[45] Whereas for the rest of his specimens, including parts of the male reproductive system such as the testicles, Ruysch gives a description in both Latin and Dutch, in the case of preparations of the female genitalia the descriptions are only in Latin and mostly marked with an asterisk.[46] This suggests that these preparations were not meant for the eyes of the common man: he would, after all, not be able to decipher the Latin text. In Ruysch's crowded cabinets, these preparations were situated behind other preparations or on the top shelves.[47] This apparent modesty was rather selective though, as it were only the female genitals that were exclusively described in Latin. The inner female reproductive system, the wombs and ovaries, did get a Dutch description.

Seen in context, Ruysch' cautious though rather transparent attempt to chasten his *Thesaurus* can be quite easily understood. Ruysch and some other Amsterdam physicians were also responsible for educating the city midwives, and were called upon when the midwives were confronted with cases they did not know how to deal with. Judging by a controversy that arose from Ruysch's involvement in such a difficult birth, discussions of the nether regions of the female anatomy were both the object of fierce skirmishing between surgeons and physicians, as well as the topic of gossip and laughter for the common man. It all started with an anonymously published 1677 satire attacking Ruysch and two other obstetricians in their alleged maltreatment of a woman suffering from a severely ruptured perineum, followed by a pamphlet attacking Ruysch

43 Also see Laqueur, *Making Sex*.

44 Ruysch, *Ontleedkundige Verhandelingen*.

45 Knoeff, "Sex in Public," 48, 51.

46 Ruysch, *Thesauri Anatomica*, vol. II, 44, no. 18, vol. III, 20, no. 15, vol. IV, 17, no. 62, vol. V, 35, no. 70.

47 Knoeff, "Sex in Public," 46–51.

published by a certain Paulus Pijl, who, according to a small book in defence of Ruysch, was actually a certain Jan Coendering. Pijl/Coendering's most remarkable accusation was that Ruysch was "trespassing on the domain of women."[48] This illustrates that it was still a commonly heard idea outside the medical faculties that the secrets of the female body and the practice of obstetrics should remain exclusively among women and in the hands of midwives.[49] No wonder Ruysch was cautious when it came to describing and displaying preparations of female genitals.

The Ruysch preparation kept in Leiden of the hand with the vagina hanging from its perky bow-tie would likely have been considered a bit racy by eighteenth-century visitors to an anatomical collection, even though Ruysch himself might not have thought it to be offensive. This is one of those preparations where the addition of lace and ribbons does not really seem to make it any less horrid. The fact that it is not described in Ruysch's *Thesaurus* and ended up in Leiden suggests that it may have been a gift—wanton or otherwise—to one of his Leiden friends, who also probably did not keep it on display for all his visitors to see. In fact, its history in Leiden is rather unclear.[50] But as Albinus studied with Ruysch, and because anatomy professors made the trip from Leiden to Amsterdam and vice versa frequently, there is a fair chance Albinus saw this preparation in Ruysch's collection.[51] There he must also have seen the other preparations in which Ruysch combined hands with human tissue, naturalia, and exotica, and took them as an inspiration for his own preparation of a hand holding a choroid membrane: a classier and more allegorical combination than that of hand and vagina.

In addition, Albinus' preparation refers to a discussion about the structure of the eye in which Ruysch had an important part. This becomes clear from a short chapter in Albinus' *Academicarum Annotationem*, a collection of notes on his research that he published towards the end of his life (1754–1768), when his sight started to fail him and at a time when most of his preparations must have been made already.[52] In four pages, Albinus explains that repeated dissection and injection of the membranes of the eye has led him to the conclusion that the membrane named Tunica Ruyschiana by Ruysch is actually nothing more than a part, a lamina, of the choroid membrane already described by oth-

---

48 Anonymous, "De Koeckoecx-Zangh." This controversy is discussed in more length in Kooijmans, *Death Defied*, 109–32.

49 Also see Park, *Secrets of Women*, 117–20.

50 I have not found any references to this preparation dated before 1952.

51 Kooijmans, *De Doodskunstenaar*, 363–5.

52 Punt, *B.S. Albinus*, 11.

ers before Ruysch. However, Albinus quickly states that Ruysch should still be acknowledged for discerning this particular part of the choroid membrane.[53] What Albinus first mentioned in writing in 1768 he had probably been telling his students for years when showing them the preparation of the child's hand with the choroid membrane.

By creating a preparation that referred both the anatomist's two most important senses and to a contemporary research topic—the anatomy of the eye—Albinus made a preparation that confirmed his reputation as one of the great anatomists of his time. Moreover, with the delicate removal of skin and nails from the arm and the intricate preparation of the eye membrane, Albinus grasped the opportunity to both place himself within the tradition of Ruysch, and to distinguish himself with his novel research, historical awareness, good taste, refinement and elegance. In these preparations we thus see the various aspects of aesthesis clearly reflected: they were made in pursuit of gaining and sharing knowledge about the human body, through a process that is highly tacit and relies on sensory perceptions and experience. The result had to be a pinnacle of anatomical skill and perfection, as elegant as possible, and was a commodity: a stable, movable and even tradable object. That the aspiration to elegance and perfection not only served the purpose of showing the intricacies of perfect human anatomy becomes clear if the textile additions to these preparations are taken into account.

Faith and Fashion: Lace

The lace-rimmed hands clearly have multiple meanings, yet the exact purpose of the sleeves varies somewhat. De Roemer has touched upon Ruysch's visual and textual references to the 'embroidery' and 'fabric' of the human body in his 2009 article. As embroidery was a virtuous, industrious craft that often used valuable materials like gold thread and silk, it was something an anatomist would gladly associate himself with.[54] A reference by Ruysch to Psalm 139 indicates that he thought of his preparations as combinations of man-made and God-made fabric.[55] Ruysch indeed created preparations in which pieces

53 B.S. Albinus, *Academicarum Annotationum*, 1768, vol. VII, chapter IV, 39–42. Leonard and Gray, *Gray's Anatomy*, 183: In fact, Ruysch would indeed get credited for his discovery: now normally referred to as the ciliary vessels, the *Tunica Ruyschiana* is still mentioned in brackets in the 2005 pocket edition of *Gray's Anatomy*.

54 De Roemer, "From Vanitas to Veneration," 11–13.

55 Ibid., 12.

of preserved human tissue are clear stand-ins for embroidery, such as when he puts a square of it in the hand of the skeleton of a foetus that holds it up to its eyes as a handkerchief. But the iconographic function of the handkerchiefs held by the skeletons is different from that of the sleeves and collars covering stumps. For example, whereas the first is natural human tissue referring to, or representing man-made tissue, the latter is artificial tissue not representing or referring to human tissue. These covers are apparently just that, covers, and they only occur in 'natural' places, such as around wrists, ankles, and necks. Genitals or internal organs are sometimes suspended from, but never decorated with lace.

Moreover, lace is something different than embroidery, and the sleeves of the two preparations under discussion here are not embroidered, but rimmed with strips of lace. The most straightforward explanation for the presence of lace trimmings and ribbons would be that they are simply additional signs of elegant anatomy. They certainly fulfil a role in this respect. After all, elegance has the twofold meaning of precision and refinement on the one hand and good taste on the other. The production of lace, though a fairly common activity in an early eighteenth-century household, certainly requires precision and refinement, just like the creation of the injected anatomical preparations these laces are adorning.[56] In addition, the 'sleeves' on these preparations are of course actually a kind of napkin bandage; wrapped around the part where the hand or arm has been severed from the rest of the body, thus covering up a probably rather unsightly stump. Hence they fulfil a purpose that is an essential part of aesthesis: they regulate the core disgust evoked by the sight of severed body parts by covering up the most vivid reminder of that separation.

Ruysch was not secretive about using decorations to avoid disgust at all. In his *Thesaurus Anatomicus* he wrote that he had tried "to improve the horrible sight of dissected and severed parts with adequate decorations to such an extent that they would not discomfort the eyes, nor cause any scares or disgust," and "to take away the disgust, as man is naturally afraid of dead bodies and their parts."[57] This is a tactic that is found in preparations of human anatomy during the entire eighteenth century, although by the last decades the covers

56 Kraatz, *Lace. History and Fashion*, 55–9, 187–8.

57 Ruysch quoted in Driessen-Van het Reve, "Frederik Ruysch in Rusland," 127: "... om 't afschuwelijk gezicht der ontlede en afgesnede delen met welgepaste cierzelen zodanig te verbeteren, dat het de ogen niet mishage, nog enige schrik en walging veroorzaken zal (...)." Ruysch, *Alle Werken*, 1000: "Ik doe het om den mensch alle afkeer te benemen, die dog van nature anders gewoon is een schrik te hebben voor dode menschen, en hare delen."

are mostly crude linen bandages instead of these lacy affairs. Moreover, any moral disgust aroused by the sight of a child's severed limb in a phial might be tempered by a lace-rimmed sleeve. It hints at a certain modesty and sophistication, and it seems to have done the trick for most of the visitors to the Ruysch collections. The young German lawyer Zacharias Conrad von Uffenbach for example, visited in 1711 and describes the preparations as 'delicate' and 'beautiful' in his travel diary.[58]

However, there is more to these lace-rimmed sleeves than their instrumentality in defying disgust. Although they certainly were the result of aesthesis, ideas about elegance and beauty were undeniably influenced by fashion. This already shows from the numbers. Albinus' sleeved arm was the only lace-decorated preparation among the three hundred-plus preparations in his collection, whereas Ruysch's preparation 4070-13 is one of fifty-four lace-rimmed preparations of human anatomy in the St. Petersburg Kunstkamera—still only about seven percent of the total if we take the current, digitized collection as representative, but significantly more than the sole Albinus preparation with lace. There are a lot of arms, hands, and feet, mainly of infants and young children, and even a number of heads, all adorned with sometimes rather elaborate lace sleeves, trousers, hats, and collars. This is less surprising when viewed against the background of the fashion of the second half of the seventeenth and the first half of the eighteenth centuries. The lace on the preparations is so-called bobbin lace, which was relatively easy to make. The technique allowed only for rather simple patterns and fairly small strips of lace to be made, but was popular in all of Europe for centuries because of its light, frothy appearance.

Interestingly, men and children also wore profuse lace adornments on the dark, gloomy clothing fashionable in the late seventeenth century, at least in equal measure to women. In some areas, lace was even primarily a sign of masculinity. This explains the profuse lace decorations on Ruysch's preparations: they do not just cover up unsightly stumps, they also commodify the preparations into fashionable, desirable objects. Around 1720, exactly when Albinus was coming of age, this started to change.[59] Although men would continue to wear lace cravats and cuffs for much of the eighteenth century, they were much more subdued and a far cry from the elaborate adornments of the seventeenth century—and that is reflected in anatomical preparations, where we find no lace decorations whatsoever after the 1720s. Now it was the women who wore much more lace, as its qualities, such as suppleness and softness, became more and more associated with femininity. Eventually, the new

58 Von Uffenbach, *Merckwürdige Reisen*, vol. 3, 384, 640.

59 Kraatz, *Lace. History and Fashion*, 34–107.

simplicity that became fashionable with the advent of Romanticism by the late eighteenth century meant that lace almost completely disappeared from fashion.[60] These developments are reflected in both anatomist portraits and their preparations: in the course of the century, both anatomists and preparations become dressed more soberly.

This poses one more question. Was the use of lace in anatomical preparations a particular early eighteenth-century Dutch trait? Ruysch's collection is the earliest in which so many wet preparations have been preserved; as Ruysch was one of the godfathers of wet anatomical preparation techniques there simply is nothing from the same period to compare it to. In Albinus' collection, we see that the use of lace is already an exception. The only remaining anatomical collection from the same period which is more or less comparable to the Albinus collection in terms of numbers and the skill with which it was assembled is that of the London-based Scottish anatomist John Hunter. But although these two men were contemporaries and skilled anatomists, Hunter's collection contains no such thing as lace-rimmed preparations. In the context of political history, this makes perfect sense though. England was never much of a centre of lace production, and importing and even wearing lace was forbidden in England for much of the eighteenth century in attempts to protect the country's own cloth production.[61] Moreover, a collection like Clemente Susini's late eighteenth-century wax anatomical Venuses reflect the decline of lace as an important fashion element in the second half of the eighteenth century.[62] Although the waxes have been donned with real hair and pearl necklaces, and are resting on pink and white satin cushions rimmed with tassels, not a strip of lace is to be found here.

Although in the Ruysch preparations that had embroidery-like tissue in them, such as the skeletons carrying handkerchiefs, there were religious and symbolic connotations, the practice of wrapping stumps in 'natural' places with lace-rimmed sleeves and collars had primarily another aim and was strongly influenced by contemporary fashion. The use of 'sleeves' and covers as such continued throughout the eighteenth century and served the same purpose all along: it was meant first and foremost to take away the source of much possible visceral disgust in the people who saw these preparations. Yet the addition of lace rims reflects local fashion, and with that, changing ideas

60 Ibid., 34–107.

61 Earnshaw, *Lace in Fashion*.

62 For the decline of lace in fashion see Kraatz, *Lace. History and Fashion*, 80–2. For more information on wax modelling and Clemente Susini, see i.e. Maerker, "Model Experts," *Model Experts*, Messbarger, *The Lady Anatomist*.

on elegance and beauty. The covers of stumps in the Leiden collections, first decorated with lace, but later on made of plain fabric, hid the proof of what the spectator still knows is there, but at least is not confronted with directly: that these are severed body parts of dead people.

Botanical Symbols: Plants in Preparations

Ruysch's and Albinus' aesthesis was decisive for the way their preparations looked. Understanding the preparations as the result of anatomical investigations and demonstration of anatomical and physiological knowledge is relatively easy. However, understanding their symbolic and natural philosophical meaning, and the tacit contemporary knowledge they represent, requires more in-depth investigation and interpretation. This is also true for the other two preparations discussed in this chapter: the ones in which human anatomy is combined with a plant. The preparation in which the human specimen has been reduced to a pile of white flakes on the bottom of the flask is the more arresting of the two. What were those flakes before, and why is the preparation still in a storage room? This preparation was another creation of Albinus; it unfortunately fell apart somewhere in the past six decades. It is (or maybe I should say was) described in the 1945 catalogue as follows:

> Entire epidermis of a child's hand, removed from the hand like a glove. Tied and hung from a sprig of *Aster africanus*. (…) Unique and wonderful preparation![63]

Once again, we are confronted with an anatomical masterpiece, morbid but beautiful. Once again, pressing questions remain unanswered by textual description of the preparation. Why did Albinus make a separate preparation of the epidermis and why did he hang it from a sprig of *Aster africanus* or, as we know it, purple groundsel, and not simply from a horsehair, as was common? According to Pole's anatomical handbook, preparations of the cuticle

---

63 Transcription from the card system in the LUMC Anatomical Museum: "Volledige opperhuid van een kinderhand, als een handschoen van de hand verwijderd. Dichtgebonden en opgehangen aan een takje Aster africanus. Helaas is de opperhuid en het proximale deel doorgescheurd. Overigens uniek en prachtig praeparaat!" Bijzonderheden: *Aster africanus* zie: Leon Plukenes Phytographia part. 1, tab. 15, fig. 4, 1769. Ingeschreven door J. Dankmeijer; 1945. Sandifort, E.: 1793: Museum Anatomicum I: Acad. Lugd. Bat.: pag.29: beschrijving. Epidermis manus infantis, Ethanol 80%."

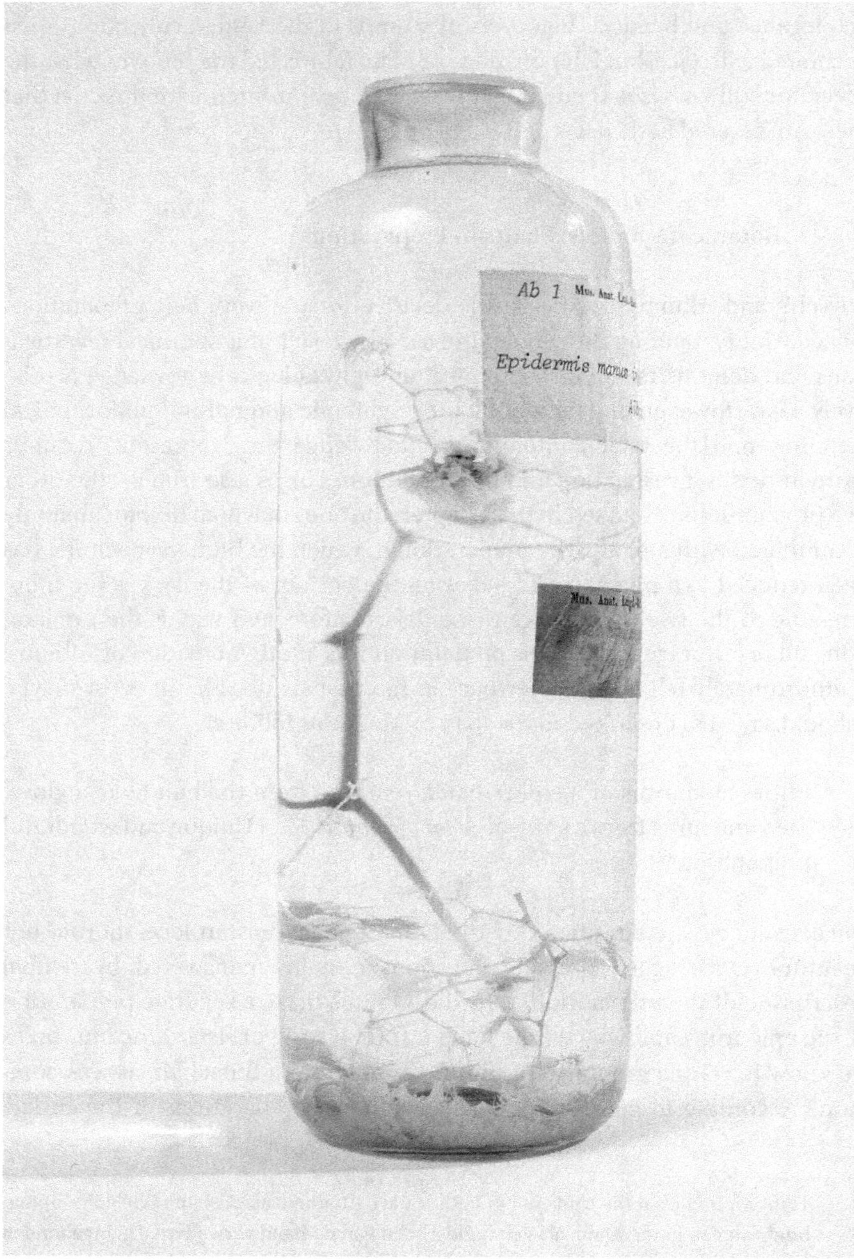

ILLUSTRATION 4.3 *LUMC Ab0001. B.S. Albinus (1692–1770). Entire epidermis of a child's hand, removed from the hand like a glove. Tied and hung from a sprig of Aster Africanus. (Epidermis now on bottom of phial.) © Anatomisch Museum LUMC 2012. Photographer: Arno Massee.*

of the hand or foot were so light that they barely needed suspension; when suspended, Pole warned, they would easily tear when the glass was turned.[64] Albinus' teacher Ruysch made at least two glove-like preparations of the skin of a hand as well. He writes of three preparations made of one hand: the hand itself, and two preparations of layers of skin. One was a preparation that not only contained the skin of the hand, but also an exotic oily plant from Ceylon, and its 'long yellow fruit.'[65] The other must have been miniscule, as it was the epidermis of the hand of a four-month old foetus, similar to the preparation Albinus mentioned in his letter to Nesbitt. The preparation is only mentioned briefly in Ruysch' *Thesaurus*, and appears not to have been attached to a plant.[66] It appears that for Ruysch the main purpose of these preparations was to show his skills and knowledge of the world.[67] He may have had some additional iconographic motive as well: taking off one's gloves and handing them over to someone else was traditionally a symbolic gesture of parting with one's possessions, and in seventeenth-century portrait painting, holding luxurious gloves was a symbol of power and riches.[68]

Although some of Albinus' preparations appear to resemble those made earlier by Ruysch, there are important differences between the two men and their work. Both Ruysch's work and personality were marked by the large scale, the grand gesture, a certain drive to amaze and entertain. He liked to brag about how he was a man of practice, not much interested in books.[69] Albinus, in contrast, although he did publish an anatomical atlas and also had a big

---

64 Pole, *The Anatomical Instructor*, 97–8.

65 Ruysch *Thesauri*, vol. 9, 9, no. 62: "Phiala, in qua continentur in liquore Foetus humani brachium, vitae specimen repraesentans; huic a latere adsita est ejusdem manus epidermis & corpus reticulare cheirothecam apprime repraesentans, ne unguibus quidem ademptis. Hocce objectum adspicientium oculos allicit, & visui gratiam conciliat. Plantula apposita est balsamina, mas Zeylan. Fructu longo flavo."

66 Ruysch, *Thesauri*, vol. 3, no. 46: "LVI: Een vlesje met heldere vogt waarin te zien is het Opper-velleke van een handje van een onvoldrage vrugtje van ontrent 4 Maanden dragts, 't welk een kleen Handschoentje komt te verbeelden."

67 Pole, *The Anatomical Instructor*, 97–8: mentions that a glove- or sock-like preparation of the hand or foot of a child is not that hard to make: after a period of maceration of the body part in water, the cuticle can supposedly be easily pulled off, washed out and suspended. Yet Pole wrote a century after Ruysch first made these preparations, and from the admiring comments of visitors and students it appears this was not commonly known among anatomists in the early eighteenth century.

68 See i.e. http://www.christies.com/features/2010-june-the-subtle-iconography-of-georg-pencz-782-1.aspx, http://www.dutchpaintings.com/moreelse.html.

69 Ruysch, *Thesauri*, vol. 3. 46.

collection in which he showed off his refined anatomical skills, was much more private and cautious. He certainly did not publish popular tour guides to his collections—in fact, he housed his collection separated from the collection in the anatomical theatre, away from the general public- nor did he create any preparations that were aimed primarily at communicating vanitas and showing of his skills, like Ruysch sometimes did.[70] If Albinus made a preparation Ruysch-style, he did so because it conveyed symbols, showed his skills and contributed to his anatomical and physiological investigations. With Ruysch the anatomical importance of his preparations sometimes appears to be no more than a footnote, whereas with Albinus, there is more balance between refined presentation and conveying knowledge. So although the spectacular idea of a glove-like preparation of the hand may have come from Ruysch, Albinus only used it because it fitted in with his research and his quest for beauty and elegance in equal measure.

As noted before, much of Albinus' research was concerned with the sensory organs, including the skin. In order to find a satisfying answer to the question of why this elaborate skin preparation was made, we have to study Albinus' theory and research of the nervous system and the sensory organs more thoroughly. Although Albinus never extensively published on physiology, he did perform anatomical experiments to answer physiological questions, about which he gave lectures. Hence we know he was primarily concerned with the physiological subjects of the 'liquidum nerveum' and the functioning of the nerves and sensory organs, particularly the skin. Through experimenting, mainly with injections, he eventually reached the conclusion that there was no such thing as a 'liquidum nerveum,' and that the 'animus' or spirits were the cause of voluntary motion. The seat of the spirits, he argued, was the cerebrum. However, he failed to identify the material transmission of spirit to the muscles and thus the cause of automatic movements.[71] Still, Albinus thought that the sense of touch was likely to be the most closely connected to the nervous system, as this sense can only discern the tactile qualities of an object when it is in direct contact with the body. This led him in to discuss the sense of touch before any other external sense after he discussed the brain and the nerves in his lectures.

According to Albinus, the sense of touch was thus divided into two classes: the internal and external sense. The animus or spirit, memory and the will moved the internal sense of touch. The external sense of touch was moved through the impulses received by the external sensory organs of skin, tongue,

70 For the housing of the Albinus collection see Huisman, *The Finger of God*, 108.
71 Punt, *B.S. Albinus*, 81.

nose, eyes, and ears.[72] The skin or epidermis, as well as the nails and hairs connected to it, were a continuing source of fascination to Albinus. His elaborate skin preparations showed that the skin consists of different layers, and in his lectures he used the glove-like preparation of the skin of the hand to argue that the upper layer of the skin was actually insensible, just as the nails, and that the external sense of touch was situated in the lower layers of the skin.[73] This explains why Albinus chose to remove the epidermis and nails from the child's hand in the first preparation and to make the second one. What this does not explain though, is why he chose to make the second preparation the way he did: with a piece of plant for suspension.

Collecting plant species, both local and exotic, was very popular in the seventeenth and eighteenth centuries. Plants were commodified, both used in medicine and sold as rarities to the rich for their ornamental gardens.[74] From the eighteenth century onwards, the creation of taxonomic systems became increasingly important. Plants could simultaneously be rarities, symbols, and drugs. Combining preparations of human anatomy with pieces of plants thus fitted in with the early eighteenth-century epistemic culture of anatomy, and it was something Albinus must also have picked up from his teacher. Ruysch, who had started his career as an apothecary and was therefore familiar with many kinds of indigenous and exotic plants, frequently used sprigs of plants as suspension and support in preparations. At first sight, Ruysch's motives for doing so seem purely practical. For example, in his description of a preparation of a piece of negroid skin, he states that the twig he used in it serves to push the preparation against the glass of the phial:

> Phial filled with liquid, in which are part of the epidermis and Malpighian body of a negro, and to which is added (...) a twig of the rarest and smallest thistle of Alpinus, (which presses the object against the glass).[75]

---

72 Box 1742, vol. II, 282: "De Tactu ... hois dividitur in 2 Classem, alii interni, alii externi, internis pendentes ab animo, et memoria et voluntas, externis, q p [quam posset] organe externe afficintur, apparet ssa [sensa] interna obstinere ad nervos, d. [dixi(t)] externa ..."

73 Box 1742, vol. II, 292: "Demonstrabo nunc chirothecam epidermidis detractam integre con communibus (=coib) pilis atque unguibus. Praeteream noto hanc epidermidem nullam habere sensae tactum, nam ubi densa se potest deradi sine dolore." In the margin Box has described the preparations Albinus used to illustrate his argument. Here it reads: 'Praepar. Chiroteca Epidermidis de fotu [ganens],' which translates as 'Preparation of a glove of epidermis detracted through heating.'

74 See i.e. Cook, *Matters of Exchange*, chapter 8.

75 Rusych, *Thesauri*, vol. 10, 1–2: "VI. Phiala liquore repleta, & in ea portio epidermidis & corporis reticularis Malphicii, ex Aethiopissa: & ut illa melius in conspectum venirent, ramum rarissimi cardui minimi Alpini (qui objectum ad vitri parietem premit) addidi."

Yet why would Ruysch mention so explicitly that the twig was of 'the smallest thistle of Alpinus'? He used that particular plant more than once as well, as appears from a description and drawing of one of his preparations published after his death: here a "bird from Ceylon" is positioned on "the smallest thistle of Prosp. Alpinus."[76] For the learned man of the seventeenth century there would have been no doubt: the repeated use of the smallest thistle, described by the Venetian Prosper Alpinus (1553–1617) in his book on the healing plants of the African continent, was a reference to the African, or more generally exotic, origins of these specimens.[77] De Medicina Aegyptiorum was first published in Venice in 1591, but was reprinted well into the eighteenth century. It was a standard work that a wealthy former apothecary like Ruysch very likely knew and of which he probably even kept a copy in his personal library.

Similarly, Albinus' choice of a sprig of *Aster africanus* as a suspension was certainly no coincidence. This becomes clear when we look up the plant in contemporary botanical handbooks, such as William Curtis' 1792 *The Botanical Magazine*. He lists the synonyms for *Aster africanus* and starts his article on it with the words:

> Linneaus has given to this charming annual the name of elegans, on account of the great beauty of its flowers, the florets of the radius being of a most brilliant purple, and those of the disk bright yellow.[78]

Linneaus (1707–1778) was a contemporary of Albinus, and the two met when Linneaus visited Leiden, where he published his 1735 *Systema Naturae*.[79] In his quest for beauty and elegance, Albinus chose a sprig of *Aster africanus*, alias *Senecio elegans*, the plant with the beautiful flowers, to mount this exceptional preparation. The glove-like epidermis of the hand, with its delicate, transparent fingers flaring out like the elongated, pointy petals of *Senecio elegans*, hung in its phial like a beautiful flower on a stem. If anything, this preparation is the materialization of eighteenth-century aesthesis: it is both the result of the practical hands-on investigation of the senses, something which can only be done through sensory perception, and of seeking beauty and perfection in

76 Ruysch, *Alle Werken*, 1276: "De I. Fig. van de I. Tafel. Verbeelt ons een droog glas, waar in een Ceylons vogeltje gevonden werd, zittende op de kleenste distel van Prosp. Alpinus."

77 Alpinus, *Medicina Aegyptiorum*, 180.

78 Curtis, *The Botanical Magazine*, vol. 7.

79 That Linnaeus and Albinus met appears from letters like one written by Albrecht von Haller to Linnaeus while he was in Leiden, asking him to give Albinus a sample of a plant. See Von Haller, 14 April 1737, *The Linnaean correspondence*, linnaeus.c18.net, letter L0182.

nature and sensory experience. The latter is largely tacit and to be found in the materiality of the preparation itself: only the informed contemporary observer would notice and understand the reference to the beauty of the flowers and the name of *Senecio elegans*.

From Show-Pieces to Obscurity and Back

Certain types of preparations enjoyed popularity throughout the eighteenth century, such as delicate epidermis preparations. In the 1799 catalogue of the Middelburg physician and collector Evert van Visvliet for example, we also find a 'sock-like' preparation of the foot of a child.[80] The addition of naturalia and artificialia to preparations of human anatomy quickly disappeared though. The informed contemporary observer must indeed have understood and appreciated the references made through the use of lace and plants. However, it appears such references quickly became irrelevant, even if the preparation itself was still appreciated. More than a century after Ruysch sold his collection to tsar Peter the Great, preparations with similar decorations were no longer made, but some still found their way into the Leiden anatomical collections. These preparations appear to have been valued in varying ways by the late eighteenth and early nineteenth-century keepers of the Leiden anatomical collections, father and son Eduard and Gerard Sandifort. Eduard Sandifort (1742–1814) was appointed as praelector of anatomy and surgery in October 1770 and shortly afterwards, in May 1771, was extraordinary professor for these subjects.[81] After the purchase of the Albinus collection in February 1772, he was made ordinary professor and ordered by the university curators to write a report on how the anatomical collections should be housed and used.[82] He was also asked to make a catalogue of the anatomical collections, including the recently acquired Albinus collection. This catalogue, with an extensive foreword, appeared in two volumes in 1793. Sandifort's son Gerard (1779–1748), appointed professor of anatomy in 1801, published a third volume covering the most important acquisitions since 1793, namely the collections of Bonn and Brugmans, which appeared in 1827. From the way father and son described the collections, it appears the importance and understanding of aesthesis in anatomy was diminishing after the turn of the century. Yet as will be seen from the following chapters, it is too easy to say aesthesis was replaced by a separate

80 Visvliet, "Catalogus Eener Uitmuntende Verzameling," 176, no. 17.
81 Molhuysen, *Bronnen*, vol. VI, 66, 71.
82 Ibid., p. 78.

realm of the aesthetic, or that the quest for beauty disappeared from anatomical discourse entirely.

Eduard Sandifort, in his preface to the first two volumes of the Museum Anatomicum in 1793, notes that the collection of Wouter van Doeveren contained a number of Ruysch preparations, and although he does not mark them out in the list of preparations, he is keen on using appreciative adjectives in his descriptions. 'Nitidissimi,' meaning very elegant or pretty, and 'perfectissime' are among his favourites. Interestingly, a glove-like preparation like the one Albinus described to Nesbitt, and that Ruysch described shortly in his Thesaurus as "the epidermis of the hand of a premature foetus of about four months carriage, which depicts a small glove," seems to resurface in the Van Doeveren collection as "Epidermis both perfectly intact and stripped from the cubit, from the hand of a four month old fetus."[83] Another preparation in the Van Doeveren collection, that is likely to be a preparation made or inspired by Ruysch, is the hand of a newborn child, stripped of the epidermis, holding a piece of wax-injected placenta.[84] In iconography and preparation technique, this specimen resembles the Ruysch collections much more than any of the things van Doeveren had made himself.[85] So even though Sandifort did not individually identify these preparations as possibly having been made by Ruysch, he did acknowledge their presence in his foreword, and described them with appreciation.

Something similar goes for the Albinus collections. The Albinus cabinet was purchased by Leiden University after Albinus' death in 1770 for a substantial sum, and extra cabinets were ordered for the display of the preparations by his successors, his brother Frederik Bernhard and Eduard Sandifort. More than twenty years after his death, the Albinus collection was still praised by Sandifort senior as a treasure. Sandifort recounted how the twenty-two year old Albinus lectured on anatomy in perfect Latin, illustrating his lectures with beautiful anatomical preparations that 'did not even scare the most sensitive

83 Ruysch, *Thesauri*, vol. III, 46: "LVI: Een vlesje met heldere vogt waarin te zien is het Oppervelleke van een handje van een onvoldrage vrugtje van ontrent 4 Maanden dragts, 't welk een kleen Handschoentje komt te verbeelden." Sandifort, *Museum Anatomicum*, vol. I, 104, no. XCIV: "Epidermis, ab utroque cubito atque manu foetus quadrimestris perfectissime detracta & illaesa."

84 Sandifort, *Museum Anatomicum*, vol. I, 103: "LXXIV. Cutibus & manus extrema infantis recens nati, a quibus, detracta epidermide, cutis rubet, maculis adinstar exanthematum apparentibus. Parte cutis ablata conspiciuntur musculi. Manus tenet portionem placentae, cujus vasa optime sunt impleta."

85 See chapter 5 for more preparations by Van Doeveren.

onlookers.'[86] Sandifort was also very protective of the Leiden anatomical collections, in particular the Albinus preparations. According to the travel diary of the English botanist and founder of the London Linnean Society James Edward Smith, who visited Leiden in 1786, Sandifort was even ridiculously overprotective. Smith wrote:

> Professor Sandifort shewed me the Anatomical Theatre, and the preparations of Albinus; the latter can be seen in his presence only. Among them are some fine things, particularly the preparations relating to the progress of ossification in the foetus, a favourite subject of Albinus; but on the whole this collection will bear no comparison with either of the Hunterian Museums. In it are a few pieces prepared by Ruysch, so celebrated in his time, which are so bad, that no London anatomist would deign to keep them.[87]

From the early nineteenth century, the attitude towards the old Albinus and Ruysch preparations appears to have changed quickly in Leiden too. Preparations made by or in the tradition of Ruysch and Albinus were still incorporated in the collections in the early nineteenth century, but now the reasons for doing so appeared to have to do less with appreciation for particular predecessors and their skills and methods, and more with practical considerations. For example, a preparation of a child's arm decorated in Ruyschian style was probably included in the nineteenth-century Leiden anatomical collections because it was an easy and harmless way to get a good example of a severe smallpox case. The preparation of the arm and hand of a child is covered in smallpox marks, injected with red wax, decorated with a lace-rimmed sleeve, and accompanied by two sprigs of a plant.[88] It came with a part of the cabinet of the deceased Amsterdam anatomy professor Andreas Bonn (1738–1818), bought by Leiden University in 1822.[89] There is a fair possibility that this is actually a Ruysch preparation: the abundant pox marks make it unlikely that this is an Albinus preparation and the type of lace used on the sleeve is, as we

86 Sandifort, *Museum Anatomicum*, vol. I, XXIV.

87 Smith, *A Sketch of a Tour*, vol. 1, 15.

88 G. Sandifort, *Museum Anatomicum*, vol. III, 354: "CCCXXXVI. Brachium dextrum infantis, qui variolis laboravit; cernuntur recentes cicatrices variolarum. Arteriae sunt impletae unde parum rubet cutis."

89 According to, Elshout, *Het Leidse kabinet*, 88, the Bonn cabinet was purchased in 1819, but the university archives only contain documents that register the purchase in the year 1822 (AC2/228 uitgaand 1822, AC2/162–180: No. 161; 22 okt / 25 okt., No. 107, 2 November 1822).

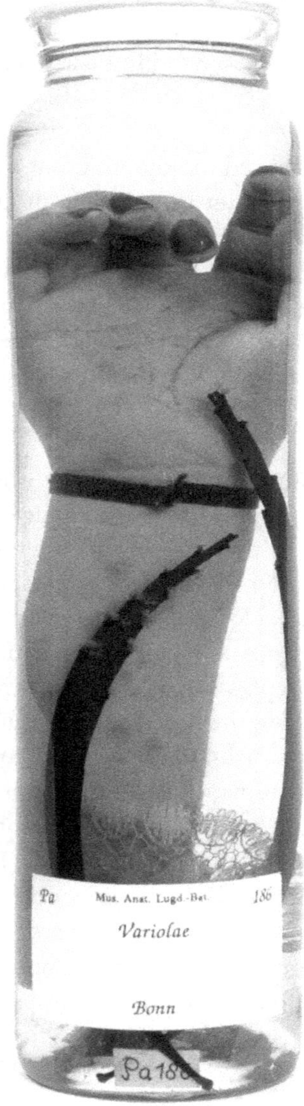

ILLUSTRATION 4.4 *LUMC Pa0186. A. Bonn (1738–1817). Preparation of the arm and hand of a child, covered in smallpox marks, injected with red wax, decorated with a lace-rimmed sleeve and accompanied by two sprigs of a plant.*
© Anatomisch Museum LUMC 2012.

have seen previously in this chapter, typical for the early eighteenth century. Finally, the iconography and materiality of the combination of wax injections, lace-rimmed sleeve, and plants point in Ruysch's direction.

Obviously Bonn could not have known Ruysch personally (Rusych died seven years before Bonn was born), but as they both spent most of their career in the same Amsterdam anatomical circles, it is not unlikely a preparation by Ruysch was at some point given to, or bought by, Bonn. Another possibility is that the preparation was in the collection of the Amsterdam Surgeon's Company, the seller of the Bonn collection, and was for some reason listed as part of that collection. We will probably never know, as there are few records left of the transaction. Either way, the preparation ended up in Leiden in 1822 through the purchase of a part of the Bonn collection. According to Gerard Sandifort, the University did not purchase the entire collection, as not all of the preparations were useful. In the preface he states:

> ... of anatomical preparations, only those should be kept which were destroyed or damaged by time or accidents in other collections, and which are missed, and moreover, those which add to an already present series of preparations and which are considered useful to clarify science; equally from the pathological preparations only those preparations were kept that added to our collection and showed the traces of disease in parts of the body.[90]

The rest of the Bonn preparations, physiology and pathology specimens, mainly of diseased bones, were shipped off to Ghent in what is now Belgium. In the listing of preparations in the *Museum Anatomicum*, the preparation is described as "The right arm of an infant, who battled the pox; freshly wasted scars are to be seen. Arteries are injected red and so can be seen a little in the skin." So the importance of this preparation for the Leiden collection becomes clear: it was about the relatively fresh pox scars that could be seen on it. No mention is made of the lace-rimmed sleeve and the pieces of plant that accompanied this specimen in its phial. Although it is not unlikely that at least some learned men in the early nineteenth century still easily grasped the meaning of such additions, they were not deemed worth mentioning—or removing, for that matter. In the course of the nineteenth century, the appreciation of the scientific community for elaborately decorated preparations such as Ruysch's would only diminish further.[91] New ideas about scientific standards rendered

90 G. Sandifort, *Museum Anatomicum*, vol. III, 2.

91 Hendriksen, Huistra and Knoeff, "Recycling Anatomical Preparations."

them archaic and obsolete, something discussed by Hieke Huistra elsewhere.[92] His work did remain a source of inspiration for artists and writers though, as an Italian dialogue on the experience of dying between 'Ruysch and his death' from 1827, and a 2008 Dutch novel inspired by Ruysch and his work, show.[93]

Nowadays, the Ruysch and Albinus preparations are once again considered to be top pieces by their home institutions, the St. Petersburg Kunstkamera and the Leiden University Anatomical Museum and Museum Boerhaave. Recent events have made it clear that these collections will always thread the fine line between elegance and disgust, but it is exactly that slightly ambivalent nature which makes them such fascinating material.

From this material and contextual analysis of the Albinus and Ruysch preparations it appears that aesthesis pervaded all their work. On the one hand, without continually explicitly stating so but with subtle visual references, Ruysch and Albinus continuously sought to express beauty and elegance in their work, fighting possible responses of disgust with elegance and demonstrations of skill. On the other, they both strongly believed that anatomy and physiology can only be known through hands-on experience, experiment and sensory perception. To some extent these epistemic parameters led to similar visual features in their preparations: practical skill and sense of style were used to cover nasty stumps with fashion-conscious, 'naturalistic,' and elegant lace-rimmed sleeves; exotic plants were used as suspension and reference to the sophistication and erudition of the makers; and body parts were combined in such a way that an extra layer of meaning was added to the preparations.

Yet although closely connected in their material and visual language, the preparations of Ruysch and Albinus are also distinctly different in certain respects. Both anatomists used their practical skills and sense of perfection to create stable, tradable commodities from otherwise perishable, evanescent human bodies, but Ruysch did this in a much more mercantile spirit than Albinus. Aesthesis led Ruysch to publish and make preparations even of subjects that were potentially scandalous, such as the external female genitalia. For Albinus, it meant studying nervous impulses and the nervous system extensively—which to him included the sensory organs. The differences in decoration and display between the Ruysch and Albinus preparations can partly be explained by their respective personalities, but more importantly they reflect changing fashions, both in epistemology and society more generally. In the case of the lace decorations the changes are very literally related to

92 For more on this topic see Huistra, "Preparations On The Move."

93 Leopardi, "Dialogi Di Frederico Ruysch," 163–71, Rascha Peper, *Vingers van Marsepein*.

popular dress; when it comes to the use of plants as symbols it is more about epistemic fashion. Although aesthesis as such, the quest for perfection and the emphasis on practice and sensory experience, did not change much, ideas of what was beautiful and elegant did. These changes can help explain why aesthesis was rather stable throughout the long eighteenth century, but that the objects that resulted from it can seem so vastly different.

The close reading of the materiality of these preparations and their reception history has also confronted us with another pressing issue. In order to truly grasp the beauty of these preparations, we have to get an idea of what beauty was to the eyes of their original beholders. That can only be done through combining an analysis of contemporary discourse with studying their materiality. The fact that a precious preparation, such as that glove-like hand, has deteriorated beyond repair in the past decades, in spite of huge curatorial efforts to maintain objects like these, is a vivid reminder of the reality that these preparations will not last forever. Fewer will remain over the years, making it increasingly difficult to retrieve and understand their original context and meanings. Although no image can ever replace the physical object, having images of disappeared objects is better than not having any visual documentation of them at all. Rather than letting historical anatomical collections quietly wither away, we should try to document them and learn from them while we can. The virtual Ruysch museum project of the Kunstkamera is a wonderful example of how this can be done.

Finally, notwithstanding their turbulent reception history and the ongoing moral and conservation issues these collections will stir up, the specimens that now constitute the eighteenth-century Leiden anatomical collections tell an extraordinary tale: that of a combined quest for beauty and perfection in anatomical and physiological knowledge through experiment and observation. These preparations are the result of aesthesis, which combined observation and hands-on experiment with a largely tacit search for beauty and perfection. By analysing their materiality, that knowledge can be accessed once again, in a way that would be impossible with text as our only source. As will appear in the subsequent chapters, although aesthesis would play an important part in the making of anatomical preparations in Leiden throughout the entire eighteenth century, the idea of what was a beautiful, perfect, elegant preparation certainly did change.

CHAPTER 5

Beautiful Monsters: How Deformity Can Be Elegant

So far, the preparations from the eighteenth-century Leiden anatomical collections appear to be predominantly examples of normal—or even perfect—human anatomy, prepared in such a way that they convey the aesthesis of anatomy. The result of particular yet tacit ideas of beauty, perfection and elegance, most of the body parts were chosen for these preparations because they were already in themselves perfect specimens: there are no obvious pathologies, and most of them were (part of) young, lean, healthy bodies. This made it easier for the creator of the preparation to inject and preserve them elegantly, skillfully distracting the attention of his audience away from the disgust and horror associated with severed body parts. Yet ultimately, severed body parts on the brink of decay—the ultimate emblem of disgust—are the constituents of this collection, and as in the course of the eighteenth-century research into pathology and abnormalities became increasingly important, an aesthesis ex negativo emerged: an aesthesis of the ugly and the imperfect.

In this chapter an analysis of two seemingly unrelated preparations shows that the mostly tacit quest for beauty and perfection by the makers of the eighteenth-century Leiden anatomical collections could simultaneously offer a way to deal with the ugly, the deformed, and the monstrous, and that their aesthesis-based approach included elements—such as poetry—we would not recognize as 'scientific' nowadays. The first is a preparation of a deformed dog with a cleft lip taken from the collection of professor Wouter van Doeveren (1730–1783), the second is a preparation of an ear from the apparently perfect collection of B.S. Albinus (1697–1770), which was maintained and invested with new meaning by Van Doeveren in the 1770s.[1] Analysing the materiality of these preparations clarifies the first as being a representative of the complicated and changing relation between humans and animals in the eighteenth century, and, studied through the combined perspectives of aesthesis and modern aesthetic theory, it will clarify the role of anatomical preparations in general and that of monstrous specimens in particular. The second preparation will turn out to be a signifier of a major eighteenth-century medical controversy and social issue, unveiling an unexpected intermingling of academia and aesthesis in settling controversies. Yet before exploring the preparations in detail, some points regarding the relevant theoretical framework need to be

1 LUMC catalogue numbers of the dog and the ear are Ad0022 and Ab0100 respectively.

 | DOI 10.1163/9789004262775_006

noted. Moreover, this chapter will show that, even by the end of the eighteenth century, looking at, and the making of, anatomical preparations, like drawing, remained simultaneous acts of seeking beauty, selection, and accentuation.[2]

Physical mutilation and deformation and the disgust associated with them have occupied philosophers past and present. William Ian Miller in 1997 wrote:

> Deformity and ugliness are (…) unsettling because they are disordering; they undo the complacency that comes with disattendability; they force us to look and notice, or to suffer self-consciousness about not looking or not not looking. They introduce alarm and anxiety by virtue of their power to horrify and disgust.[3]

This seemingly inextricable connection between deformity and a state of alarm led Winfried Menninghaus to state in his 2003 philosophical work on disgust that aesthetics and anatomy are opposing disciplines, as 'the aesthetically pleasing body has no interior, hence allows no dissection or anatomy.'[4] Yet in his explorations of theories of disgust over the past 250 years, he readily admits that 'the field of aesthetics always covers more ground than just the beautiful', and discusses the work of eighteenth-century authors such as Burke, Mendelssohn and Lessing, all of whom address in their writings on beauty the paradoxical pleasure we find in (representations of) horror, the ugly, gruesome, and revolting.[5]

As we will see in this chapter, these paradoxes are less contradictory than they first appear. Philosopher Carolyn Korsmeyer's recent take on that eighteenth-century aesthetic experience par excellence, the sublime, will prove to be extremely helpful in understanding the aesthesis at work in the creation and appreciation of these preparations.[6] This is the kind of anachronistic conceptualism Nick Jardine advocates and that helps to clarify the historical meaning and significance of knowledge, objects, and events beyond the significance attached to them in the period when they occurred.[7] But the leads are once again given by the preparations.

2 Daston and Gallison, *Objectivity*, 104.

3 Miller, *The Anatomy of Disgust*, 82.

4 Menninghaus, Disgust, 55.

5 Ibid., 33–6.

6 Korsmeyer, "Fear and Disgust," *Savoring Disgust*.

7 Jardine, "Uses and Abuses," 253.

Monsters: A Dog with a Harelip

Preparation Ad0022 from the collection of Wouter van Doeveren is a little spotted dog with what is commonly known as a harelip or cleft lip. In a way, this preparation is an exception among the preparations that serve as case studies for the collections in this book. It is the only preparation so far that has not been injected, dissected, or decorated. Obviously, there has still been some sort of intervention from the collector: he has chosen this specimen over others to include in his collection, and has carefully preserved it in conservation liquid and chosen an appropriately shaped and sized phial to display it. Yet, it is remarkable in its plainness and the lack of artful enhancement in the form of coloured waxes, mercury, plants, lace, beads, or coins. In fact, I will argue that in this case the absence of such additions and the plain materiality of the preserved dog answer the questions it evokes just as well as do the added materials in the other preparations. In this section, I will explore this plain deformed dog and try to answer the issues it raises just as I did with the more thoroughly manipulated preparations previously discussed. So why did Van Doeveren collect preparations of congenital deformities, and why did he want this particular preparation, a dog with a harelip for his collections?

Van Doeveren's reasons for wanting a dog with a harelip in his collection while he was primarily interested in human anatomy may initially seem obscure. Of the 441 preparations in the Van Doeveren collection listed in the 1793 Museum Anatomicum, only fifteen are of animals, but four of those preparations were dogs. The others were four pigs, three chickens, a kitten, and a pigeon. Obviously, these are all domestic animals and thus relatively easy to acquire. The most likely reason for the dominance of dogs and pigs in the small animal section is that their anatomy and teratology are fairly comparable to that of humans. Both were traditionally used as stand-ins in human anatomy research, but a mature dog of average size is easier to handle than a fully grown pig, which can weigh up to two hundred kilograms.[8] Moreover, a pig is potential food, whereas although dogs were also primarily kept as working animals such as hounds, guard or cart dogs until the nineteenth century, they were not

8 Porter (ed), *Cambridge Illustrated History of Medicine*, 62: In Antiquity, when dissecting humans was forbidden, the Roman physician Galen (CE 129–216) already taught human anatomy based on the dissection of apes, pigs and dogs. It should be noted that 'comparative anatomy' was always primarily about human anatomy—animals initially served solely as stand-ins that clarified human anatomy, and only from the eighteenth century onwards did animal dissections also start to serve as investigations into animal anatomy.

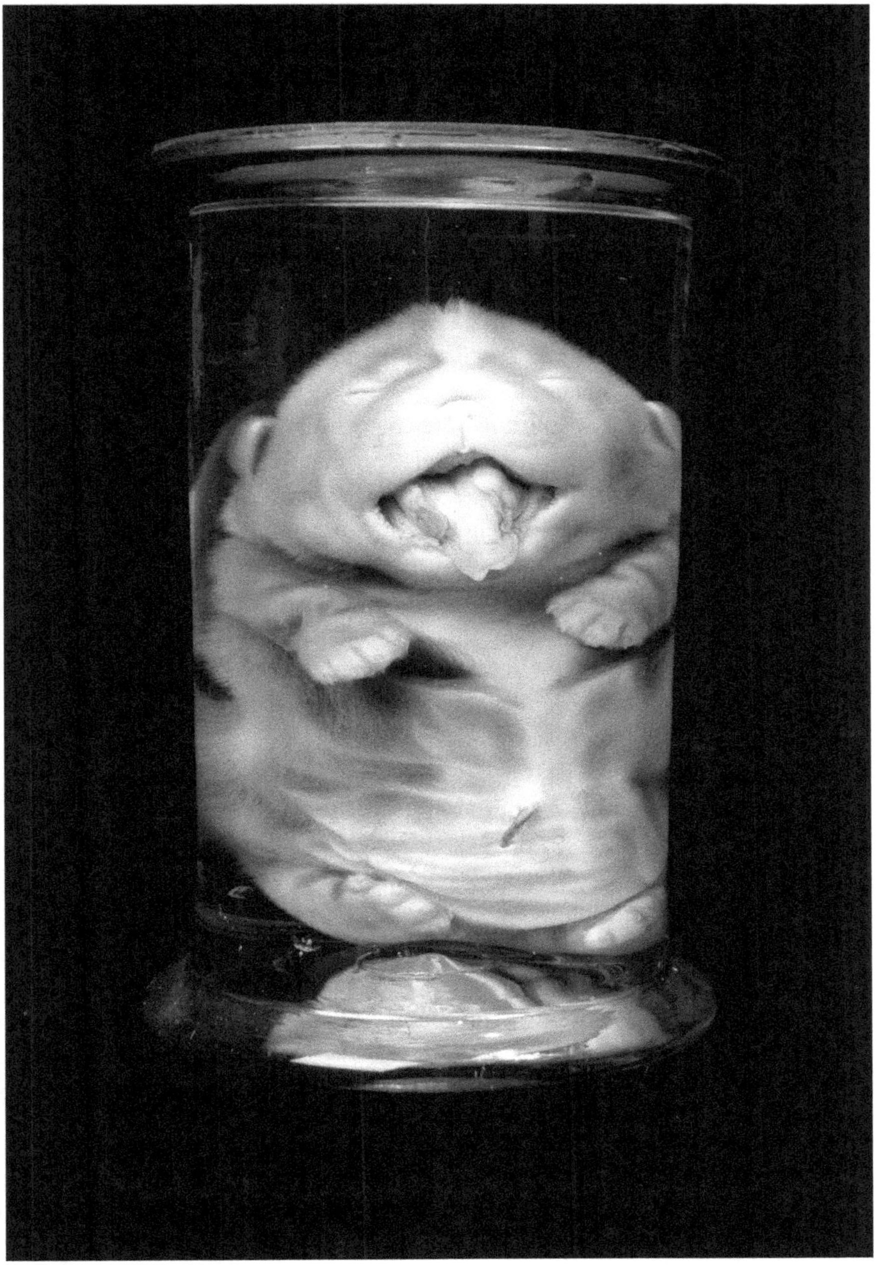

ILLUSTRATION 5.0 *LUMC Ad0022. Wouter van Doeveren (1730–1783). Dog with cleft palate.*
© Anatomisch Museum LUMC 2012.

considered appropriate food.[9] Being inedible, dogs were considered appropriate subjects, and very useful in medicine and anatomy.

For example, Van Doeveren's contemporary Martinus Houttuyn (1720–1798), another medical doctor with a Leiden degree, published an extensive multiple-volume natural history based on Linneaus' system in the second half of the eighteenth century. In the second part of the first volume, he discussed the origins, races and uses of dogs. Although the eating of dogs was something Houttuyn dismissed as only customary in other parts of the world, according to his texts dogs could well be used in the manufacturing of medicines. 'Young Dog Balm,' a kind of 'eucalyptus' balm made out of young dogs boiled in olive oil with herbs, was available in most city apothecaries, and purified dog fat was used to make a kind of chest relief balm. Small dogs, Houttuyn wrote, could be used as living hot-water bags to treat colic pains with their body heat, and having them lick wounds was generally known to speed the healing process. Whereas these latter uses seem rather innocent, it was also perfectly acceptable to have a dog lick the legs of a gout sufferer—although it was likely to kill the dog, it cured the sufferer.[10] Socks of dog leather could be used to suppress bouts of gout, gloves of dog skin were beneficial for contractures in the joints of the hand, and the skin of a freshly slaughtered dog could be applied as a curative to any body part affected by atrophy.[11] This must have been of interest to Van Doeveren, who was a long-time gout sufferer himself and eventually died of the disease.

In anatomy research, dogs were used as cheap replacements in dissections and for experiments on living animals that were not allowed on humans.[12] As Houttuyn put it:

> The Art of Dissection has taken great Advantage of the Dog. It has served Asellius in the discovery of the Lymphatic Ducts; Pecquetus in the finding of the Lymphatic Tube that runs through the Chest; Harvaeus to demonstrate the Circulation of the Blood (...). Yes, this Animal makes the

9 Zijlmans, *Hond & Baas*.

10 Houttuyn, *Natuurlyke Historie*, 66–9.

11 Jütte, "Die Haut Als Heilmittel," 167. Before the eighteenth century, human skin was preferred over dog skin for these purposes, but as this was scarce and expensive, dog skin was a good alternative.

12 Guerrini, *Experimenting with Humans and Animals*.

> foremost Subject of Comparative Anatomy, and for that reason the Dog has since long bore the name of Martyr of the Physicians.[13]

Although the lap dog was quickly gaining popularity with the upper classes in the late eighteenth century, no one would have been shocked by the use of dogs in vivisections and dissections—structural anti-vivisectionism first appeared in the course of the nineteenth century.[14] Confusing to the modern mind, one use of dogs did not exclude the other: Descartes is known to have experimented on dogs and yet he had a pet dog that he liked, and did not vivisect or dissect it as far as we know.[15] Van Doeveren too experimented on dogs. The earliest records of this are the irritability experiments he performed on dogs while studying in Paris with his friend Jan Tak.[16] Experiments defining irritability and sensibility, (the responses of muscles and nerves to stimuli), were essential in mapping the separate and combined functioning of the sensory organs, nerves, and brain. Physiologists throughout the eighteenth century did similar experiments, ever since Albrecht von Haller had started them in the 1730s.[17] The main subjects in these experiments were dogs, for the reasons already stated: their anatomy and physiology resembles that of humans, they were widely available and relatively easy to handle.

Bearing the above in mind, the dog in Van Doeveren's collection does not seem so awkward anymore. Dogs were used abundantly in eighteenth-century anatomy and medicine, so Van Doeveren was probably given this dog by a colleague or patient who knew he would be interested in it as an addition to his collection of monsters, the first systematic collection of birth defects compiled in Leiden. This is why the dog mattered primarily to Van Doeveren: as his collection shows, he was not interested in a specimen of a dog with normal anatomy. He accepted only a limited number of animal specimens in his collection,

---

13 Houttuyn, *Natuurlyke Historie*, 65. The original Dutch text reads: "De Nuttigheid, welke de Ontleedkunde van den Hond getrokken heeft, is groot. Hy heeft ASELLIUS gediend tot het ontdekken van de Chylvoerende Vaten; PECQUETUS in het vinden van de Chylbuis, die door de Borst loopt; HARVAEUS om den Omloop van het Bloed aan te tonen (...). Ja dit Dier maakt het voornaamste Onderwerp uit van de Vergelykende Ontleedkunde, en om die reden heeft de Hond al lang den naam gedragen van Martelaar der Geneeskundigen." (Translation mine).

14 On the rise of Dutch anti-vivisectionism in the late nineteenth-century, see Kluveld, *Reis door de Hel*, chapter 2.

15 Guerrini, *Experimenting with Humans and Animals*, 33.

16 Elshout, *Het Leidse kabinet*, 76.

17 Steinke, *Irritating Experiments*, 101, 119.

when they could show pathologies similar to the human but for which human specimens were hard to acquire. This brings us to the second question that this preparation evokes: why did Van Doeveren include a preparation of a harelip in his collection?

The first thing that draws our attention in this preparation is the dog's mouth: its upper lip is split right up to its nose. It has a cleft lip, colloquially known as a harelip, in earlier centuries sometimes also referred to as wolf mouth or devil's bite.[18] This congenital deformity of the upper lip and/or palate nowadays has an incidence of about 1 in every 700 births, meaning it is something every experienced obstetrician or midwife is familiar with. This incidence was probably higher in the eighteenth century, as women now routinely take folic acid around conception, which reduces the chances of cleft palate.[19] The correction of a cleft lip, and even a cleft palate, is now a fairly easy and cheap procedure, and people born with the deformity who are operated on early enough can go on to lead reasonably healthy lives. However, back in the eighteenth century, these procedures were obviously not widely available, nor were they very sophisticated.

Leiden professor of anatomy and surgery Antony Nuck mentioned in his 1692 surgical and medical handbook that a harelip, if not too severe, could only be corrected in young children. The procedure, described by Nuck in a rather wry tone, must have been traumatizing. The obviously unanaesthetised child would be trussed and held by an assistant, while the surgeon had to cut, stretch, and pinch together the lip with thin needles, which would only be taken out one by one in the week after the surgery.[20] The consequences of being born with a cleft lip and/or palate in the eighteenth century thus ranged from infant death due to feeding problems in severe cases, to trauma and considerable social stigma in milder cases.[21] Yet Van Doeveren's predecessors' collections in Leiden, notably the Albinus collection, did not contain any preparations of cleft lips that we know of. In other contemporary collections, preparations of cleft lips are also few and far between. In the eighteenth-century London Hunterian collection for example, there is only one preparation of a cleft lip; it is from an adult subject and has visibly been corrected in childhood.[22]

18 Bradt, *Slow Devon and Exmoor*, 63. Unfortunately the evidence for the synonym 'devil's bite' is only anecdotal.

19 Bianchi et al., "Environment and Genetics," Wilcox et al., "Folic Acid Supplements."

20 Nuck, *Operationes Et Experimenta Chirurgica*, 73–7.

21 For a case history of infant death because of cleft palate see Sandifort, *Observationes*, vol. I, 29.

22 RCSHC/P129, probably <1750.

Apparently, at least to Nuck and his contemporaries, a birth defect such as a harelip was a given that could sometimes be solved by surgery, but was not something worth collecting systematically in an anatomical collection. This preparation was indeed related to the rest of the Van Doeveren collection, and to contemporary ideas on birth defects.

The apparently sudden appearance of these kinds of human and animal monsters in the Leiden collections, collections thus far dominated largely by normal human anatomy, is in fact perfectly understandable from the perspective of its collector. Although so-called monsters had long been a prominent part of cabinets of curiosities and both private and institutional anatomical collections, these were first and foremost unique rarities; not examples representing certain categories of regularly occuring malformations. For example, in the catalogue of the Leiden anatomy theatre from 1687, we find that around the circumference of the theatre "a most curious Skeleton of a Man wasted with old age"—he probably suffered from a progressive bone disease—was placed alongside the skeletons and skins of all kinds of domestic and exotic animals, but there was no particular section within the theatre's collection devoted to preparations of deformations. Little changed in this display of singular monsters and rarities in Leiden until the second half of the eighteenth century, as professor Albinus was occupied with his own projects and could not be bothered with the collections displayed in the anatomical theatre.[23] Yet as Hagner has pointed out, in the course of the eighteenth century displays of monsters as singular examples of nature's capriciousness or as intriguing curiosities for the connoisseur disappeared, and monsters became subject to natural classifications and taxonomy.[24] Van Doeveren was the first Leiden anatomist to attempt to build a systematic collection of monsters, moving them out of the sphere of wonder and curiosity and into that of natural philosophy.[25] With this systematic approach, he moved monsters from the negative to the positive, although a tension remained, as the monsters were often not viable.

In order to understand why Van Doeveren collected what he did, including the harelipped dog, we need to understand his research interests and the big medical issues of his time. What is illustrative here is Van Doeveren's annoyance with the expensive purchase of the Albinus cabinet by Leiden University

23 For the minimal changes in the display in the Leiden anatomical theatre in the course of the eighteenth century, see for example the various editions of the *Catalogues of the Rarities in the Leiden Anatomy Theatre* kept in the British Library (1669–1753).

24 Hagner, "Enlightened Monsters," 179.

25 Ibid.

in the year of his appointment, 1771.[26] The source of his irritation was twofold. On the one hand, he might have been hoping to purchase some interesting preparations himself for further research, as he did when he bought a preparation from the cabinet of the late Noortwijk at an auction in August 1771. In a letter to his German friend Wagler, dated 23 August 1771, Van Doeveren writes how he has managed to acquire the 'old uterus and foetus,' which Noortwijk described in his 1743 Historia Uteri Gravidi.[27] In that book, Noortwijk defended the view that blood circulation of the placenta and the uterus were connected. Van Doeveren doubted this and wrote to Wagler shortly after the purchase saying that he looked forward to examining the preparation closely himself, suggesting he would be taking the wet preparation out of its phial and possibly even dissecting it again.[28]

On the other hand, Van Doeveren was likely to be irritated by the expensive investment the university made with the Albinus collection because it was of little value for the courses he wanted to teach as the newly appointed professor. Like Eduard Sandifort, he wanted not only to research and lecture on physiology and normal anatomy, he also wanted to prove his worth on big issues in the medical field. This ambition shows in many aspects of his work, to begin with his in doctoral thesis on worms in the intestines: in it, he argued that physicians should pay more attention to pathologia animata, or diseases caused by living organisms in the body.[29] Van Doeveren's role in the inoculation debate that will be explored in the second half of this chapter is also an example; the study of, and theorizing about 'monsters' is another. Like many of his contemporaries, Van Doeveren tried to answer the question of how monsters originate. The general explanation before the eighteenth century for monsters was that the mother saw something terrible during her pregnancy, or had acted immorally. In the sixteenth and seventeenth centuries, harelips had even been explained as proof that the mother had intercourse with the devil—the devil sometimes appeared as a hare, or with the head or ears of a hare in iconography, something which may well have had something to do with the reputed lustfulness and speed of reproduction of the animal.[30] These explanations had been mostly discarded by the second half of the eighteenth century—by physicians at least; the strength of the maternal imagination on

26 Also see chapter 4.

27 Noortwijk, *Uteri Humani Gravidi Anatome*.

28 Van Doeveren, MS UvA: UB: HSS-mag.: IV A 9: 2 [j].

29 Van Doeveren, *Dissertatio De Vermibus Intestinalibus*.

30 Hall, Dictionary, 149.

the unborn child lived on in popular belief well into the nineteenth, and even the twentieth, century.[31]

Physicians and natural philosophers had noted that not all pregnant women who had seen something terrible gave birth to monsters. Van Doeveren nonetheless still examined this, as appears from an episode he recounts in his book Specimen observationem. He asked the family of a woman who had just given birth to an anencephalic baby whether she had been badly influenced during her pregnancy, in particular by anything she saw.[32] Neither the family nor the woman remembered anything unusual. Yet, Van Doeveren noted:

> Nonetheless, the common people, blinded by prejudices, spread the rumour—which was strengthened by circulation—that fear induced by the sight of the slaughtering of a ram had caused the monstrosity.[33]

Moreover, the explanation of imaginatio materna or the mother's immoral actions as the cause of deformations required reason, morality, and imagination in the mother to impress the image of the terrible thing on the foetus, thus giving no satisfying explanation for monsters in animals and plants—which after all, have no reason, morality, or imagination. Yet although scarce, there were too many monsters to be coincidences, and moreover, the teleological idea of Nature as a system created by an almighty and infallible God in which everything had a purpose did not allow for mistakes. In line with the great taxonomical projects of the century, such as Linneaus' work, many tried to figure out whether monsters might be some kind of misunderstood category.[34] Maybe, some natural philosophers and anatomists argued, they formed a particular category and had a purpose (teleos) within the great system of nature that had so far been overlooked.[35]

The word 'purpose' is essential here. In the teleological view, purposefulness had long been equated with perfection, but not necessarily beauty, and this is exactly what explains this aesthesis ex negativo. Ancients like Xenophon

---

31 I.e. Wilson, "Eighteenth-Century 'Monsters'."

32 Van Doeveren, *Specimem Observationum Academicarum*, 47–8.

33 Ibid., 48: "Nihilominus tamen praejudiciis occaecatum vulgus famam excitavit, quae eundo quoque crescebat, terrorem, spectaculo arietis mactati productum, huic monstro causam fuisse."

34 Linneaus himself classified *Homo Monstrous* as a specific species, next to *Homo Sapiens*. Also see Wilson, "Eighteenth-Century 'Monsters'," 7–8.

35 Hagner, "Enlightened Monsters," 199–200.

and Longinus, and eighteenth-century writers like Mendelssohn and Burke, had used the example of the (apparent) ugliness inside the human body to explain that neither ugliness nor anomaly were contradictory to the perfection of nature, or to the will of God.[36] Moses Mendelssohn (1729–1786), a student of Baumgarten and Shaftesbury, in 1755 published his *Letters on the Sensations*. He stated that although he agreed with Leibniz that ours is the most perfect universe possible, not every expression of perfection equates with beauty. Some forms are perfect, as they fulfill their teleos, their goal, but still they are ugly. Mendelssohn argues that "underneath the skin terrible forms lie hidden" and that "all vessels are seemingly without order;" harmony and unity cannot be discerned.[37] Burke uses this same example of the intestines of man as perfectly suited for their goal, but ugly.[38] In the same vein, Van Doeveren believed his monsters to be perfectly suited to their goal, although he was not entirely sure what it was: to discover this was precisely what teratology aimed at.

This is a tendency that can also be found in the literature of the eighteenth century and beyond: both Goethe and Bilderdijk have written arguments in which they explain how the apparent ugliness of the inside of the body is actually beautifully purposeful, and that dissection is therefore not to be thought of as disgusting.[39] As early as 1748 the Irish painter Barry wrote:

> According to the definitions generally given, Beauty consists of unity and gradual variety; or unity, variety, and harmony. Our rule for judging of the mode and degree of this combination of variety and unity seems to be no other than that of its fitness and conformity to the designation of each species.[40]

Thus he connected beauty to perfection, in the sense that what is perfect or beautiful is to be completely fulfilling its purpose; but not necessarily to be pretty. This is exactly how Van Doeveren's aesthesis led him to interpret his monsters: as freaks of nature, deviations of the beautiful pattern of nature, but in these freaks, that same nature proved her astonishing power.[41] Such

36 Hammermeister, *The German Aesthetic Tradition*, 16.

37 Mendelssohn, *Wijsgeerige Verhandelingen*, 59.

38 Burke, *A Philosophical Enquiry*, 97.

39 Von Goethe, "Zur Morphologie," in *Goethes Werke*, section 2, vol. 8, 64. Bilderdijk, "Antwoord."

40 Barry et al., *Lectures on Painting*, 103.

41 Van Doeveren, *Specimem Observationum Academicarum*, 34: "Naturam haud raro in Hominum generatione aberrare a pulcerrimo illo ordine, quo ab ejus sapientissimo

a sense of purpose, of meaning, probably also helped to make monsters less disgusting, which may explain why the anatomized body appears to have been more acceptable within the public sphere then than it is nowadays.[42]

Buffon and Hunter, natural philosophers, anatomists, and contemporaries of Van Doeveren, also gave teleological explanations for the presence of monsters and their categories, so Van Doeveren's wish to contribute to the debate, expressed in his case studies of monsters, partly explains the presence of the harelipped dog in his collection.[43] However, one would expect that the tendency to no longer view monsters as one-off wonders or as devil's brood, but rather as purposeful parts of a larger system, would suggest that a collector like Van Doeveren would try to collect as many specimens of a particular deformity as possible. Yet Ad0022 was the only preparation of a creature with a harelip in Van Doeveren's collection. In the collection of diseased bones Sandifort described two children's skulls with cleft palates, which were the only other preparations with this deformation in the Leiden collections circa 1792.[44] The *Museum Anatomicum*, in which these preparations are listed, is primarily a catalogue with little or no explanation of the many individual preparations, and we have no known writings on this particular preparation by Van Doeveren, so the significance of this specimen within the collection can only be understood through analysis of secondary sources. First, Sandifort's work on cleft palates will be discussed to give an impression of how they were dealt with by the Leiden anatomists. Subsequently, the listing of the Van Doeveren collection in the *Museum Anatomicum* will be used to define the place of this preparation within his collection, as this is the earliest available listing of the collection as a whole.

Between 1777 and 1781, Eduard Sandifort published a series of four books of anatomical-pathological observations. He did so to register and disseminate his findings in dissections of pathologies, hoping that it would eventually lead

---

Auctore est imbuta, et saepius quoque Lusus amare, quibus majorem, magisque mirandam, suae potentiae vim commonstret, docent innumera monstrosorum Fetuum, eorumque diversissimae formae, exempla, a fide dignis Scriptoribus observata."

42 Alberti, *Morbid Curiosities*, 189, argues that "Over the last three centuries the sight of the corpse and the internal workings of the human body have increasingly been deemed disgusting" and that "While disgusting things were removed from the public sphere, that which constituted 'disgusting' expanded." However, Ruysch, *Alle Werken*, 691, mentioned a preparation of a foetus with a cleft palate and spina bifida that he would only show on request, as he did not want to frighten anyone with it. Also see Knoeff, "Sex in Public."

43 Fontes da Costa, *The Singular*, 125–31.

44 Sandifort, *Museum Anatomicum*, vol. I, 163–4, also see Sandifort, *Natuur- En Genees-Kundige Bibliotheek*, vol. IV, 29.

to a better understanding of birth defects. In the third chapter of the fourth book, Sandifort describes a case that is exceptional because it is one of the very few of the period in which a patient is described both before and after death. It also painfully illustrates the fate of children with severe birth defects at the time. A newborn girl with a complicated double cleft lip, palate, and nose was brought to Sandifort by her parents. Sandifort stressed that her parents were good Christians and that this was their eleventh child, but the first one with such a deformation, once again trying to shed the persistent belief that the imagination or morals of the mother influenced the formation of the unborn child.[45] Sandifort was aware that less complicated cases could be successfully operated on, and that these children could live well into adulthood. Yet a defect as severe as this one was impossible to cure, so Sandifort could do nothing but give a very detailed description, illuminated with a drawing of the child from three different angles.[46] A child with such a severe cleft lip and palate is difficult to feed, and at twenty-two weeks old, despite the best efforts of her parents, the girl died of malnourishment.[47] Sandifort persuaded the parents to give him the body, and carefully dissected and described the skull.[48]

Van Doeveren had also encountered his share of birth defects in his career. Over a decade before Sandifort published his account of the girl with a complicated cleft palate, Van Doeveren compared a number of monstrous births in a series on pathological specimens he observed while serving as anatomy professor at Groningen University. One of them was a child in a comparable state as the girl brought to Sandifort, but this child could not eat at all, and Van Doeveren notes that, as he could offer the parents no cure or comfort, he was forced to send them away again.[49] It is not surprising, being confronted with this kind of suffering, and unable to do anything about it, that Van Doeveren wanted to understand more about the occurrence of birth defects like cleft palate, and whether they had some kind of purpose. So precisely what kind of position did monstrous preparations have in his collection?

The monstra, often taken as representative of the Van Doeveren collection, in fact form a very small part of it: only 28 of the 441 preparations, so less than

---

45 Sandifort, *Observationes*, vol. II, liber IV, caput III, 30.

46 Ibid., 30–2, 36–8, Tab. VI.

47 Ibid., 32.

48 Ibid., 32–5, Tab. VII.

49 Van Doeveren, *Specimem Observationum Academicarum*, 48: "Sugendi facultate carentem, & difficillima, si ulla, deglutitione valentem, misellum tanto majore morti proximum, incurabilem, tristesque parentes optato solamine destitutos, dimittere fuerim coactus."

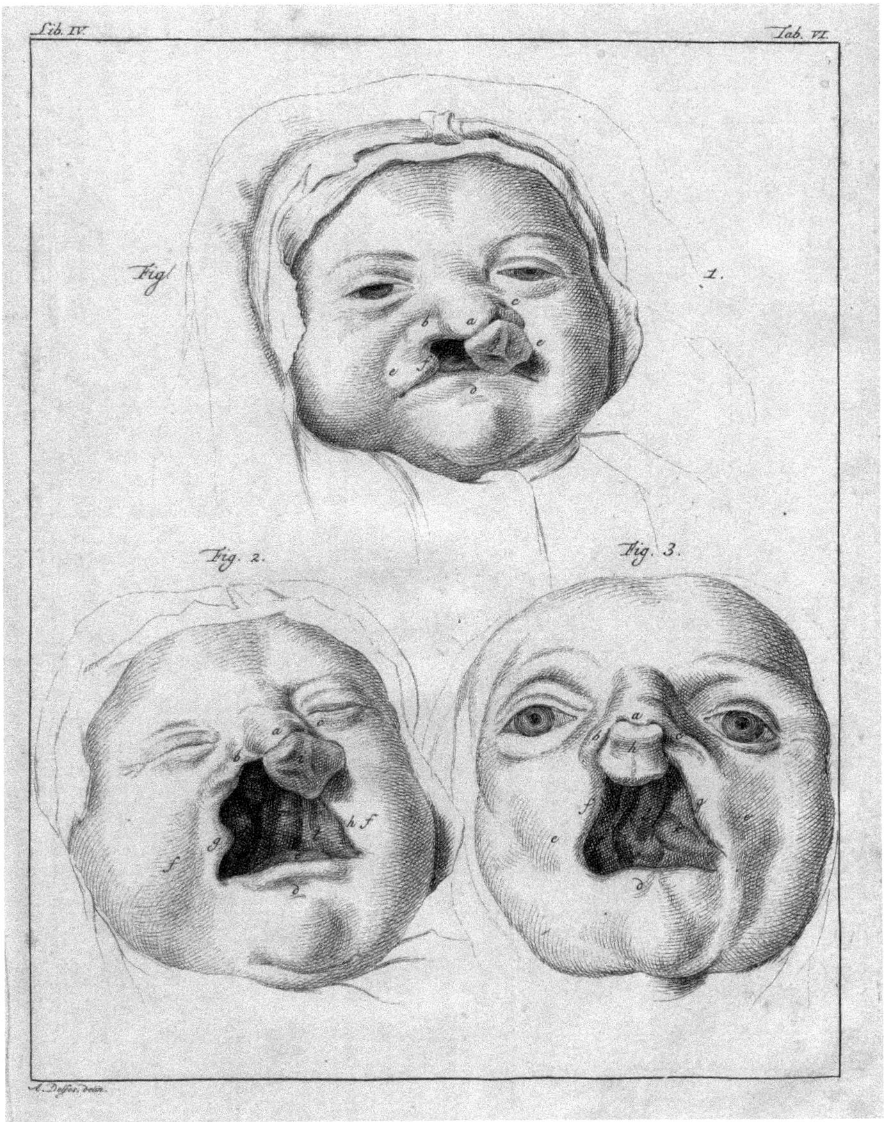

ILLUSTRATION 5.1 *Child with cleft palate, from Eduard Sandifort,* Observationes, *vol. II, liber IV, caput III, 30–2, 36–8, Tab. VI (Call number 520 C 22–23). © Leiden University Library 2012.*

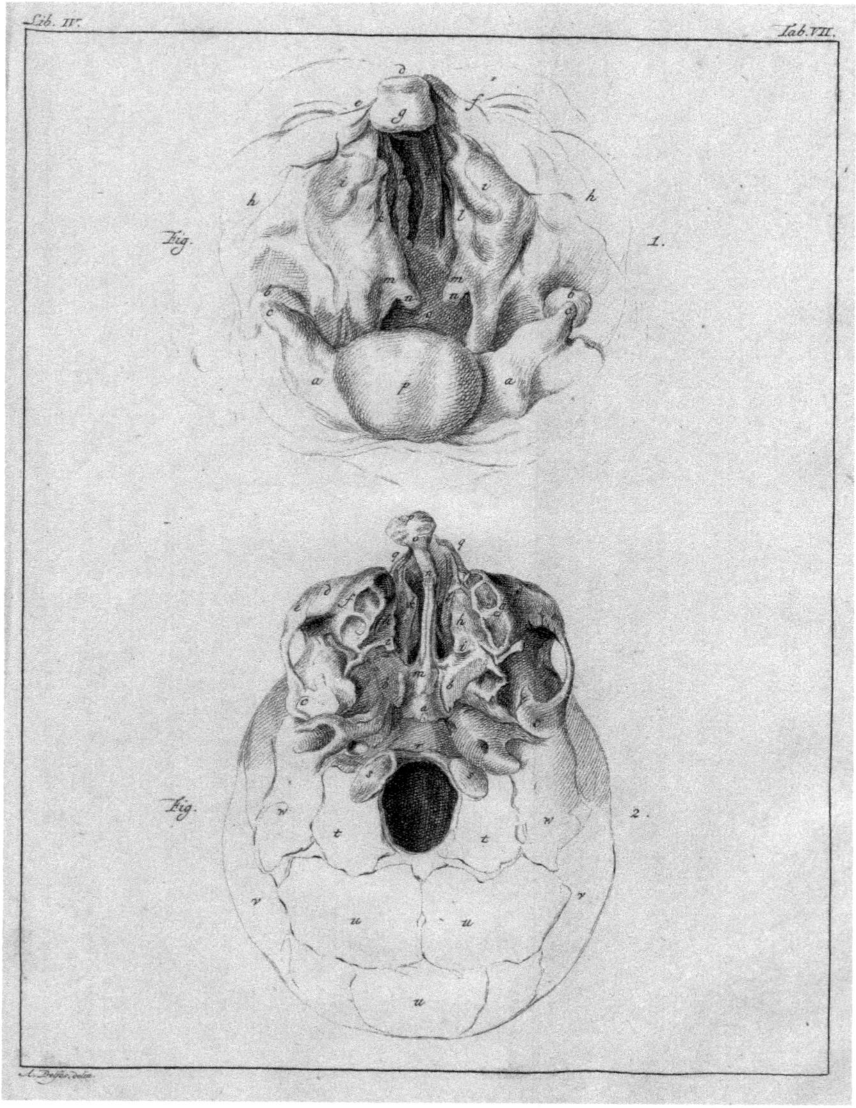

ILLUSTRATION 5.2 *Skull of a child with cleft palate, from Eduard Sandifor,* Observationes, *vol.* II, *liber* IV, *caput* III, *30–2, 36–8, Tab.* VII *(Call number 520 C 22–23). © Leiden University Library 2012.*

seven per cent. Of these monstra, 23 are human and 15 are animal.[50] As we have seen, the animal monstra are also the only animal preparations within this collection of human anatomy, suggesting they were used as additional or stand-in specimens within a category that was difficult to fill for the collector. As the debate on the function and meaning of monsters and the desire to practice taxonomy grew in the eighteenth century, monstrous specimens were likely to have become even more desirable and thus scarcer commodities for the assemblers of natural history collections. Exactly what kind of monsters did Van Doeveren collect, and what were his ideas about them? Most importantly how do Van Doeveren's monsters reflect eighteenth-century aesthesis?

When we consider the 'monstrosities' in the twenty-eight monstrous preparations in the Van Doeveren collection, we find in the human specimens section five cases of anencephaly, one specimen of what is now known as TRAP, one case of craniopagus (twins conjoined at the head), a case of spina bifida, three cases of malformed uteri, a case of hydrocephalus, a rectum with fungus, a brain tumour in a child, enlarged testicles, dactylitis ('sausage digits'), a placental tumour, a case of molar pregnancy, four parts of placentas from molar pregnancies, and a liver tumour. In the animal section, we find after the dog with the harelip a dog with five legs, one with three legs and one with two legs, a double-headed kitten, a cyclops pig, a pig with a misshapen upper jaw, two preparations of pigs with one head and two bodies, a four-legged chick, a chick with only one eye and a deformed upper beak, a chick with two heads and three legs, a grown finch with two heads, a pigeon with a split sternum, and finally a monstrous salmon head.[51]

This small, odd assembly of both congenital defects and pathologies shows that teratology and pathology were indeed emerging fields in Van Doeveren's time.[52] Also, the limited number of these preparations in his and other

---

50 Numbers based on the sales catalogue of the Van Doeveren collection (1785) and the listing of the Van Doeveren collection in the 1793 first volume of the *Museum Anatomicum*. Today very few preparations from the Van Doeveren collections remain; most were damaged beyond repair in the 1807 explosion of a gunpowder ship on the Rapenburg canal.

51 Sandifort, *Museum Anatomicum*, vol. 1, 110–11. TRAP: twin reversed arterial perfusion. A rare complication of twin pregnancies, involving an acardiac twin (i.e. a twin without a heart, and mostly also without a head and arms) whose structural defects are incompatible with life, and an otherwise normal "pump" twin. Basically, it looks like a baby with the abdomen and legs of another one growing out of its chest.

52 From the end of the eighteenth century, teratology became a particular area of interest for anatomists as there was still no satisfying explanation for the occurrence of birth defects. Many expected that a more systematic study of such deformations would lead to more knowledge about how they occurred. Moreover, disease had long been thought of

eighteenth-century collections suggests that it must have been very hard to find these specimens, let alone find more than one with a particular deformity or pathology.[53] Yet the number of human anencephales, misshapen uteri and molar pregnancies in Van Doeveren's collections suggest some kind of specialisation in these deformations. Another possibility is that Van Doeveren sought particular specimens which illustrated his theory about the function and meaning of monstrosities. This is the point where we have to turn to Van Doeveren's known writings to figure out whether this was indeed the case, and if his texts suggest anything on why he included the harelipped dog in his collection. Unfortunately, these written sources are limited to some of his correspondence and his published work, as his will stated that his personal archives with unpublished material were not to be published after his death.[54] This wish was respected, but most of these personal archives were probably subsequently destroyed in 1807, when a gunpowder ship exploded on Leiden's main canal.

Among Doeveren's first publications was a number of case studies of monsters in a book, *Specimem observationum academicarum ad monstrorum historiam, anatomen, pathologiam et artem obstetricam, praecipue spectantium* (1765), published when he was professor of anatomy at Groningen University. Although he writes nothing in it about harelips, cleft palates, or the little dog, from the illustrated case study of a double-headed lamb (a specimen not even listed in the *Museum Anatomicum*) we can learn a lot about Van Doeveren's ideas on monsters and birth defects. He was given the specimen by his Groningen colleague Nicolaas Engelhard, a professor of philosophy, and used it to discuss the problem of the generation of monsters.[55] There were generally two views on this: either monsters come into being accidentally, because of external forces, or they are so-called primigene monsters, which are actually purposeful, and thus beautiful. In line with earlier work by Von Haller and Winslow, Van Doeveren argued that this lamb proves that monsters are meant

in terms of humoral imbalances, but in this period the realization grew that disease could actually be located in very specific parts of the body, like one particular organ, limb or structure. Also see Lewis, "A Pathological Misunderstanding."

53 I.e. within the collection of Petrus Camper, http://anatomie.ub.rug.nl/index.htm, we find only one human 'monster': a specimen with a cleft palate (A0502).

54 Van der Zwaag, "Wouter van Doeveren," 66.

55 Engelhard (ed), *Feriae Aestivae Groninganae Exhibentes*: Engelhard did not publish anything on monsters, but his editorship of a book including a paper on monsters suggests he was aware of, and interested in, the debate surrounding them.

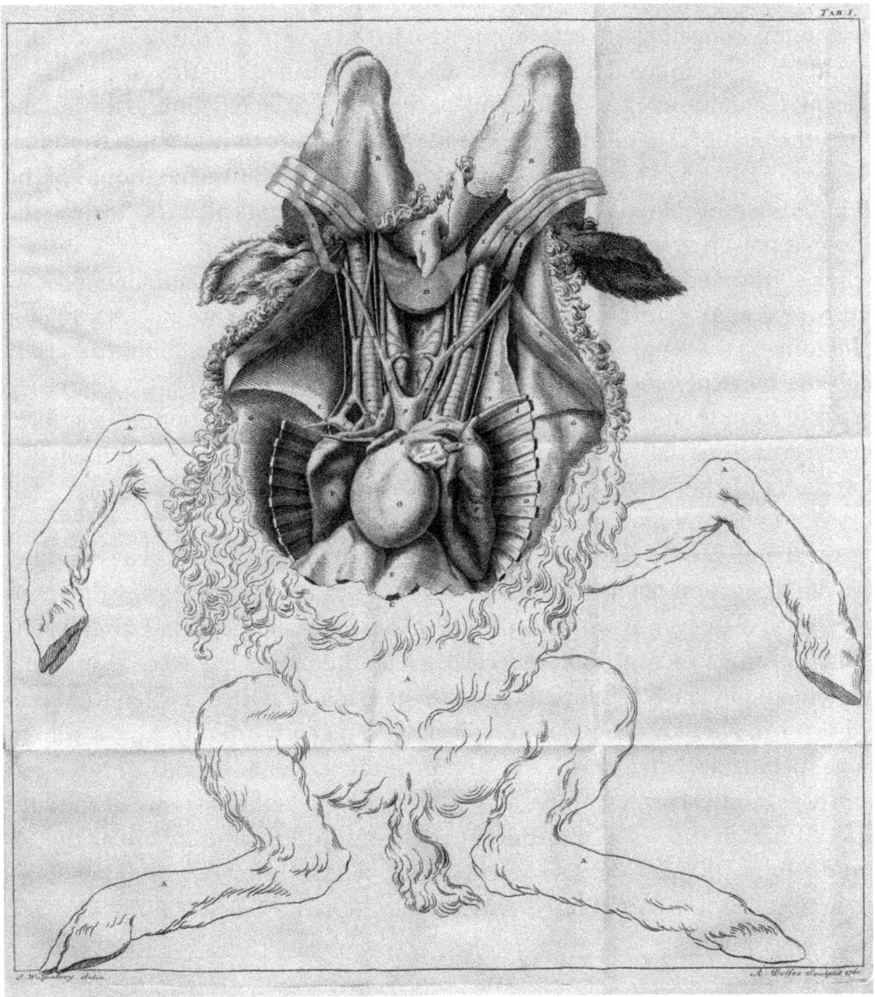

ILLUSTRATION 5.3 *Drawing of a double-headed lamb, from Van Doeveren's 1765* Specimen Observationum *(Call number 518 B 6). © Leiden University Library 2012.*

to be the way they are, because of the wonderfully symmetrical, purposeful way it was build.[56]

The lamb, according to Van Doeveren, proved that there is most likely only something he called monstra primigena, deformities which were purposefully created by nature. The possibility of monstra accidentalia, meaningless, accidental, congenital defects he rejected. In the case of the lamb, he argued, it could not have emerged from two lamb embryos pushed together, as the fusion was perfect and seamless.[57] To Van Doeveren, it was exactly in the deviations from the beautiful patterns in the Creation that Nature proved her exceptional powers.[58] This reverence for the exceptional power of Nature reminds us of the sublime, a term coined in the eighteenth century by Edmund Burke to describe the overpowering sensation of fear-inducing beauty sometimes experienced when confronted with natural phenomena, such as a ravine, a thunderstorm, or a great waterfall.[59] Interestingly, the objects of the sublime are always things or phenomena that could cause us pain, but are experienced as sublime precisely because we experience them in a situation or representation that ensures we are safe from them—think of a painting of a ravine or a panorama platform over a great waterfall.[60] In a way, that goes for these monsters as well: if alive or unpreserved, they could be dangerous for us, but dead and safely conserved in drawings and preparations, they are harmless.

Yet the eighteenth-century notion of the sublime appears to have applied only to things and phenomena much bigger than oneself, suggesting the monsters of Van Doeveren are closer to the category of eeriness than to the sublime. But the awe Van Doeveren clearly felt when studying these monstrous specimens obviously has little to do with the gross-out we seek when we watch a horror movie, for example. Korsmeyer has recently tried to find a way to both include and distinguish the kind of sublime-like awe Van Doeveren felt for these relatively small preparations. Her solution is the category of the sublate. Although the essence of the experience is in the same sphere as that of the sublime, "the sublate can come in small and subtle doses," "bringing home general truths in a particularly vivid manner," and the insights it provides "are

56 Van Doeveren, *Specimem Observationum Academicarum*, 46.

57 Ibid., 47.

58 Ibid., 34.

59 Burke, *A Philosophical Enquiry*. I have not found any eighteenth-century Dutch publications mentioning the sublime or an equivalent.

60 Guyer, "The Difficulty Of The Sublime," 35.

grasped not only with the mind but also with palpable somatic resonance.'[61] So not only can the sublate be found in small objects, such as an anatomical preparation, through its eerie beauty, a sublate object also exemplifies general truths in a sensory experience.

Korsmeyer writes with (contemporary) art in mind, yet it appears her description of the sublate serves very well in understanding the result of the aesthesis Van Doeveren sought to instill in his preparations of the apparently deviant, pathological, and monstrous. He sought beauty and perfection in the materiality of the freaks of nature he collected, relying on his own sensory perceptions, and—quite exceptional in eighteenth-century aesthesis—sometimes even voicing this explicitly, as in the case of the lamb that was so wonderfully and purposefully built, that nothing indicated the two lambs had been coincidentally pushed together.[62] Aesthesis helped Van Doeveren to deal with the disgust that always hovered so very near, using it to vividly draw attention to what he believed to be general truths. Preparations of monsters certainly had more somatic resonance for him than the drawings he made of them to illustrate his treatises, as appears from his eagerness to acquire a preparation from his deceased colleague Noortwijk of which the latter had an illustrated work.[63] From his views on monstrosity, in which Van Doeveren derived a general rule from singular examples, it appears that seeking purpose and meaning in what was generally thought to be horrific and weird was also his way of dealing with the atrocities of nature he encountered in his practices. Moreover, by preserving specimens like these in fairly non-offensive preparations and drawings, eighteenth-century anatomists made them more acceptable for display, maybe even suitably to evoking some kind of aesthetic experience, and converted them into useful and meaningful didactic instruments. Yet some controversies however could not be solved through making preparations and illustrations; in the case of smallpox inoculation, anatomists turned to poetic rhetoric.

61 Korsmeyer, *Savoring* Disgust, 134–5.

62 Van Doeveren, *Specimem Observationum Academicarum*, 43–6. On this case also see Elshout, *Het Leidse kabinet*, 81–2, Van der Zwaag, "Wouter van Doeveren", 80–1.

63 Van Doeveren, MS Uva *Letter to Petrus Camper*, 23 August 1771, UB: HSS-mag.: IV A 9: 2 [j].

ILLUSTRATION 5.4 *LUMC Ab0100. B.S. Albinus (1692–1770). Ear with smallpox scar.*
© Anatomisch Museum LUMC 2012.

Smallpox and Poetry

At first sight, preparation Ab0100 is a typical Albinus preparation: a sensory organ of average size and shape, the other sensory organ covering it, the skin, partly removed so the structure of the cutis is visible, and injected with coloured wax so it has the life-like appearance Albinus preferred. Yet it is not entirely perfect, as we might expect from an Albinus preparation: it has a tiny smallpox mark.[64] In this section, the most remarkable aspect of its materiality, the smallpox mark, so unlikely in the work of this anatomist, will be the starting point from which to explore the impact of smallpox on the Leiden anatomists, their work, and their aesthesis.[65]

Albinus' reason to inject this ear with coloured wax and preserve it in his collection was that he used it in his study and teaching of the sensory organs.[66] Yet while he tried to find perfect specimens for each of his preparations, in this case he appears to have had no choice than to accept an imperfection: a tiny smallpox mark just above the antitragus. Although the smallpox mark was visible in the illustration of this preparation in Albinus' paper on the lodging of the hair in the skin, he merely acknowledges its presence in the explanation of the table.[67] Apparently he felt no need to elaborate on this imperfection, something that seems to have confused Sandifort a little. In the latter's inventory of the Albinus collection in the 1793 *Museum Anatomicum* only two preparations of the auditory organ are listed: this particular ear as number 173 and a preparation of an ear lobe as number 174. After noting the smallpox mark on the adult human ear Sandifort lists the ear lobe with the words: "Adult human ear lobe, as in 173. But no variola scar is at hand,' as if he wanted to stress that unlike 173, this is a hard to find but perfect preparation of an ear.[68] The fact

64 I will not pay much attention to the fact that this particular preparation is an ear, as Albinus' preparations of the sensory organs have already been extensively discussed in chapter 4.

65 Staves, "The Puzzle of the Pox-Marked Body," 155: Smallpox should not be confused with pox. Pox in the eighteenth century referred to (the symptoms of) a disease that was also referred to as 'lues', 'lues venera', 'venereal disease', 'French disease', the 'loathsome disease' and 'syphillis'. Smallpox marks were condemned by some because they destroyed beauty, but were so common that they were not considered to be particularly shameful, unlike pox marks, which, especially in women, were considered a sign of loose morals.

66 For more on Albinus' study of the senses, see chapter 4.

67 Albinus, *Annot. Acad. Lib.*, vol. VI, Caput IX, 57–62, 158–9, Tab. III, Fig. IIa. In the explanation of the table, page 158, Albinus simply notes: 'cicatrix variolae;' smallpox mark.

68 Sandifort, *Museum Anatomicum*, vol. I, 152: "CLXXIV. Auricula hominis adulti, ut CLXXIII. Sed cicatrix variolae nulla adest."

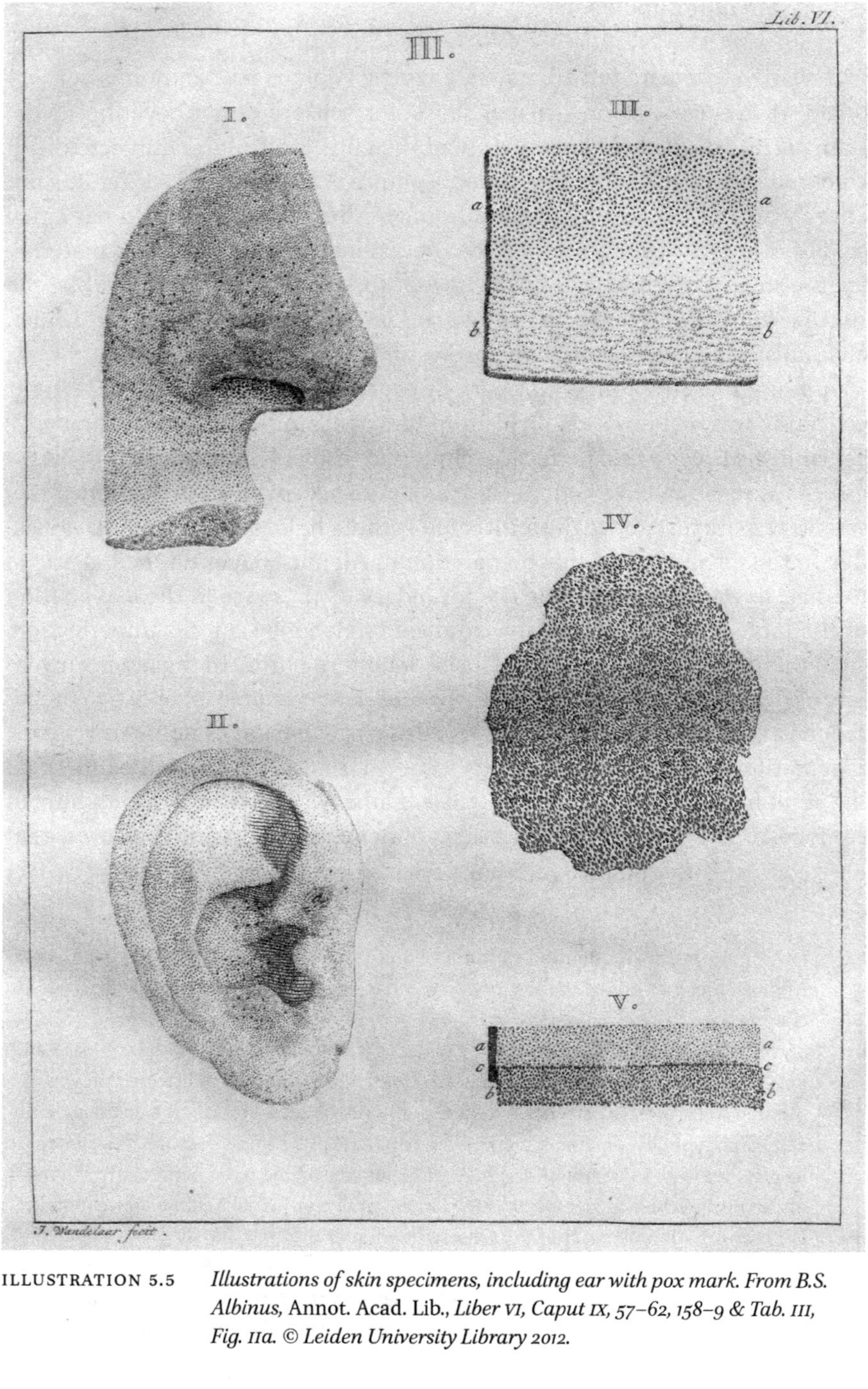

ILLUSTRATION 5.5 *Illustrations of skin specimens, including ear with pox mark. From B.S. Albinus,* Annot. Acad. Lib., *Liber VI, Caput IX, 57–62, 158–9 & Tab. III, Fig. IIa. © Leiden University Library 2012.*

that Albinus apparently readily accepted the flaw of a smallpox mark in his preparation is indicative of the widespread havoc caused by smallpox in the eighteenth century.

Every three to four years, a smallpox epidemic raged through the bigger Dutch cities. About seventy-five per cent of all people were infected at some point in their lives, and twelve to fifteen per cent of those died of smallpox. The disease was especially dangerous for the young: a child in an eighteenth-century Dutch city had a one in five chance of dying of smallpox before it reached the age of ten.[69] Those smallpox sufferers who were lucky enough to survive a severe infection were marked for life: especially on the face and hands, which could be left severely pock-pitted. With these statistics in mind, it is remarkable that there are relatively few known preparations with smallpox marks in the eighteenth-century Leiden anatomical collections: the Albinus ear is the only one of which we are certain that it was actually there in the eighteenth century. In the modern catalogue, twelve other preparations with smallpox marks are listed. One is of them listed as part of the purchase of the Bonn collection from the Amsterdam Surgeon's Guild in 1822, but as mentioned in the previous chapter, it could also be a Ruysch preparation considering its iconography.[70] Ruysch made at least one other preparation of a body covered in smallpox, but here he tried to cover up smallpox marks. He described it as follows:

> The head of a newborn of a negro, in which it should be noted that not only this head but the entire child was covered in smallpox, although the mother during her pregnancy was not affected by pox at all, and as afterwards the head was black and unsightly, to avoid all hideousness, I have decorated it all around with injected veins from another subject.[71]

From this description it appears that although smallpox marks were so common that Albinus, the Leiden champion of perfect anatomy, saw no other option than acquiescence when confronted with a specimen with a

69 Rutten, " 'De Vreselijkste Aller Harpijen'," 49–54, 82–3.

70 Pa0186, see illustration 4.4. Like in other preparations by Ruysch, in this one human tissue is also combined with lace and plants. The severity of the smallpox marks seems to exclude it as an Albinus preparation.

71 Ruysch, *Alle Werken*, 726: "Het hoofdje van een pas geboore kindje van een negerinne, waar in aan te merken staat, dat niet alleen dit hoofdje, maar ook het geheele kindje met kinderpokjes bezet waren, hoewel de moeder ten tijde hares dragts van geen pokken aangetast was geworden, en nadien het hoofdje swart en onaangezien was, zoo heb ik het zelve, om alle affschuwelijkheyd te mijden, rondom met opgevulde slagaderen van een ander voorwerp bezet."

single smallpox mark, severe cases still evoked disgust in both medical men and others.

In the Leiden collections, three other preparations with smallpox marks are first mentioned in the nineteenth-century Suringar catalogue, and another nine of unknown origins are listed. All or some of these could be from the eighteenth century too, judging by the materials and techniques used, but there are no indications they were made by Leiden anatomists, or that they were present in the Leiden collections in the eighteenth century.[72] The limited number of preparations with smallpox or smallpox marks may have various explanations. In the case of the highly infectious smallpox, there obviously is something to say for avoiding contact with them in the contagious stage of the disease, let alone going through the invasive bodily process of creating a preparation of a smallpox-infected body part.[73]

However, most eighteenth-century anatomists would have caught smallpox as a child, and it was commonly known that once you survived a smallpox infection, you were likely to be immune for the rest of your life. As almost no one had to fear infection from old smallpox marks, it is still remarkable that there are so few preparations with smallpox marks in the collections. Smallpox marks were so common that they were described as simply another feature, for example of a missing person, like their complexion, hair colour, and height.[74] Their absence can therefore only be suitably explained by the quest for elegance and beauty in preserving normal anatomy that dominated Albinus' work, and also to some extent that of his successors. Yet despite there being few pathological preparations in the Leiden collections until the late eighteenth century, such as preparations of smallpox, this did not mean that the Leiden anatomists were not involved in the debates and controversies surrounding ravaging diseases, such as the smallpox, and their possible cures.

Learned men and common people alike were appalled by the waste of life these diseases caused, and the fatalism inherent in earlier religious ideas that life was nothing but a preparation for a good death was increasingly abandoned in favour of self-help.[75] As just mentioned, it was commonly known that

72 Concerns preparations Am0089, Pa0187, Pa0277 (Suringar catalogue) and Pa0007, Pa0016, Pa0032, Pa0036, Pa0128, Pa0143, Pa0144, Pa0192, Pd0013 (origins unknown).

73 Breman and Henderson, "Diagnosis and Management of Smallpox," 1300: Smallpox was only eradicated in 1980, and because vaccination campaigns were terminated after eradication, even today virologists fear that accidental or intentional release of the stored virus could cause widespread epidemics.

74 Gowing, "Marked Bodies and Social Meanings," 141.

75 Porter, *Flesh in the Age of Reason*, 211–3.

a mild infection would prevent a person from catching the disease again. This led to practices such as putting healthy children to sleep in a bed with a child with mild smallpox, so they would almost certainly survive the disease and in future be protected from it.[76] Besides these common-sense lay treatments, inoculation was practised by medics from the eighteenth century, causing great academic and public debate. Lady Mary Wortley Montagu (1689–1762), the wife of the British ambassador to Constantinople, imported the practice of inoculation against smallpox from Turkey in 1721, where it had been common for centuries. Inoculation, also called invariolation, was the deliberate infection of an individual by making a small lesion in the skin and rubbing the fresh purulence of a smallpox patient into it.

Lady Montagu had herself been stricken with smallpox early in the eighteenth century, and probably used this personal experience to write *Saturday; the Smallpox*, a poem based on pastoral conventions in which the main character mourns the loss of her beauty to the scars of smallpox.[77] The poem is, in effect, aesthesis ex negativo: an experience of deformation, of ugliness, a sensory experience of disgust of one's own body translated into an elegant, eloquent textual account. Although Lady Montagu's children were successfully inoculated, the practice had at least as many adversaries as supporters when it was first introduced in Europe. Many considered it a risky business, which might even advance the spread of smallpox. Other objections were the financial cost of inoculation and theological concerns about interfering with God's providence.

The first recorded inoculation in the Netherlands occurred in 1748, when the French-Swiss Amsterdam-based medic Theodore Tronchin (1709–1781), who had studied in Leiden with Boerhaave, inoculated his eldest son, after the youngest had suffered a severe bout of smallpox. This case was described in 1754 by Charles Chais (1701–1785), a Walloon preacher at The Hague who wanted to refute theological objections to inoculation.[78] Other Dutch advocates of inoculation were professors Petrus Camper in Groningen, Gadso Coopmans (1746–1810) in Franeker, and Wouter van Doeveren and Eduard Sandifort in Leiden. The Europe-wide interest in the inoculation debate also emerges from the large number of entries for 'smallpox' and 'inoculation' in the commented bibliography published between 1765 and 1775 by Eduard Sandifort: no less than 85 works on these topics from all over Europe and beyond were discussed

---

76 Rutten, " 'De Vreselijkste Aller Harpijen'," 189.

77 Montagu, "Saturday; Small-Pox." Also see Spacks, *Reading Eighteenth-Century Poetry*, 90–1.

78 Chais, "Essaie Apologétique Etc.," 651–2, Van der Zwaag, "Wouter van Doeveren", 121.

by Sandifort in this decade.[79] Apart from case histories and papers we would now identify as academic, the debate was simultaneously enacted in poems and pamphlets by much the same people. The work done by, and the controversy surrounding, Leiden professor Wouter van Doeveren is illustrative here.

Wouter (sometimes anglicized as 'Walter' or latinised as 'Gualtherus') van Doeveren was born in the south-western province Zeeuws-Vlaanderen in 1730 as the son of a government official, and went to school in Goes. He enrolled as a medical student in Leiden in 1747, and took classes with professors Winter, B.S. and F.B. Albinus, Van Rooijen, Musschenbroek and Gaub. After finishing his studies in Leiden he spent some time in Paris to study with the famous obstetrician André Levret. On his return to the Low Countries in 1754, he was appointed professor of medicine at Groningen University, where he would remain for almost twenty years. Not only did he teach at the university, he also taught anatomy to the city surgeons and their apprentices, was a city physician, educated the city midwives in obstetrics, served as legal physician, researched pathology, obstetrics, surgery and inoculation, and published on all those topics.[80] In an undated letter to his friend Jan Tak he stated his support for a treatise advocating inoculation by Charles Chais (1700–1785), expressing the hope that it would lessen resistance and prejudice against inoculation.[81] Eventually, Van Doeveren was appointed professor of anatomy at Leiden University in 1771, a position he would maintain until his death in 1783. It was during his career in Leiden that the inoculation controversy in the Netherlands reached its height.

Fortunately for Van Doeveren, when he started his career in Groningen, he found himself in an atmosphere conducive to inoculation and he performed the first inoculations in 1759.[82] What began with seven inoculations in 1759 had grown into an extended inoculation program ten years later, including not only the upper classes, but also the less well off.[83] Just before his departure to Leiden, in 1770, he wrote in a letter to Eduard Sandifort:

---

79 See Sandifort, *Natuur- En Genees-Kundige Bibliotheek*, vol. 11 (register).

80 Van der Zwaag, "Wouter van Doeveren", 7–70.

81 Van Doeveren, MS UvA UB: HSS-mag.: Y 123: 0, verso II. Chais published on inoculation twice, in 1754 and 1768. The former was also published in the proceedings of the Holland Society of Sciences in Haarlem in 1754, and as Van Doeveren refers to the society before mentioning Chais, it is most likely that the letter is from 1754 or shortly afterwards.

82 Van Geuns (ed), in Van Doeveren, *Academische Redevoering*.

83 Van Doevere, *Sermo Academicus De Sanitatis*.

> It is mainly the unanimity of all the physicians in this city which has contributed to the excellent progress of inoculation, a work of so much importance: I have doubted whether there was one amongst them, who did not advocate and practice inoculation: rare example! The noble permission and public declarations of the foremost of our Theologians and Preachers; their prayers and giving thanks from the pulpit; yes, setting an example with their own children and relatives; have contributed to strenghten the work, and to have many embrace the practice.[84]

With Groningen still fresh in his memory, Van Doeveren used his inaugural lecture in Leiden, delivered on 6 May 1771, to once more advocate inoculation as a means to battle smallpox epidemics.[85] However, in the same month, the burgomasters of Leiden officially asked the medical faculty for advice on inoculation policy. The faculty replied that inoculation would not be wise when there were no smallpox cases in the city, in consideration of the risk of contamination.[86] Van Doeveren, his pride piqued, wrote a response to the faculty on 2 June. He explained that although he could understand the point of view that inoculation on a wide scale was not necessary as long as the smallpox had not been diagnosed in the city, he feared that the advice given by the faculty would be misinterpreted and used by the adversaries of inoculation to totally abandon the practice. According to Van Doeveren, once smallpox appeared either in the city itself or in neighboring towns and villages, the university and city should indeed actively promote inoculation.[87]

Initially, his plea seemed to be of little use. On 13 June, the judiciary of Leiden requested the medical faculty as well as all physicians, surgeons, and apothecaries to no longer inoculate or attempt to cure smallpox, threatening punishment if caught.[88] However, when an epidemic seemed to be nigh in late 1772, the ban was lifted and Van Doeveren started inoculation on 16 february 1773 with the six year old child of a fellow professor, treating thirty others in the following two months. Nonetheless resistance among the people of Leiden

---

84 Van Doeveren, *Brief Aan Eduard Sandifort*.

85 Van Doeveren, *Academische Redevoering*, 48.

86 Van der Zwaag, "Wouter van Doeveren," 132, Daniëls, *De Kinderpok-Inenting in Nederland*, 110–1.

87 Van Doeveren, *Academische Redevoering*, quoted in Van der Zwaag, "Wouter van Doeveren," 131–2.

88 Daniëls, *De Kinderpok-Inenting in Nederland*, 111.

was fierce, and in the course of 1773 a poetic slander campaign arose against Van Doeveren.[89]

At the centre of the controversy surrounding Van Doeveren and his plea for inoculation were two Dutch satirical poems, published anonymously. The tacit aesthesis attitude of the Leiden anatomists and their social circle was not only apparent in their dealing with the shift which would eventually lead to the new disciplines of pathology and comparative anatomy, it also appears in the controversies over anomalies they were part of. In the eighteenth century, it was very common amongst scholars to fight their intellectual battles not only in private correspondence, but, as there were no academic journals to speak of, also in public media such as books, spectatorial reviews, and printed pamphlets. In many cases, learned men (they were mostly men) even tried to settle their disputes through poetry. Poetry had long-standing status as an authoritarian mode of social and philosophic commentary, which simultaneously allowed for openly expressed feeling and the deployment of sensibility as a basis for judgment.[90] After all, according to Baumgarten, the poem is the perfect sensate discourse, the pre-eminent manner in which to express what is perceived through the senses.[91]

Moreover, neo-Latin literature flourished in the Low Countries until the 1830s, a phenomenon which has been partly ascribed to the lack of standardized grammar, vocabulary, and poetic rules in Dutch, as well as to the fact that classical philology was an integral part of every well-reared young man's education. Already at the Latin schools, between the ages of seven and fourteen, boys learned to write letters, speeches, and poems in Latin. In the Northern Netherlands, the dominant Protestant poets were mostly professors, who often specialized in the genre of the descriptive and moralizing Horatian ode.[92] Medical heroic-didactic poetry, the description of the symptoms and cures of virulent diseases, was a literary tradition originating in antiquity that had been enthusiastically cultivated since the Renaissance. It had started with Hieronymus Fracastorius' *Syphilis* in 1530 in Verona, and for many eighteenth-century scholars, new discoveries were not validated until they were celebrated in poetry.[93] The lofty art of Latin poetry was also a man-

89 This controversy has also been partly described in Dutch in Van der Zwaag, "Wouter van Doeveren," 120–44.

90 Spacks, *Reading Eighteenth-Century Poetry*, 185.

91 Baumgarten, *Reflections on Poetry*, §1–9, 37–9.

92 Guépin, "Het Bataafse Oor." 138–42.

93 Sacré, "An Imitator," 521–2.

ner in which to show one's eloquence, and to discuss what were in themselves rather gory affairs in an agreeable manner.

The poems attacking Van Doeveren were "Portrait van G. van D...... N" and "Op de inenting der Kinderen-pocken" (on the inoculation of smallpox).[94] Although not written in Latin, they were most likely the products of the pen of Van Doeveren's fellow professor, the old physician and botanist Adriaan van Rooijen (1704–1779), who was a fierce opponent of inoculation.[95] These works were more squibs than satire. On one hand, the Netherlands did not have much of a tradition of satirical poetry, especially not compared to England. On the other, the writers of these rhymes ignored many traditions of the classical satire. Most importantly, instead of a critique of the vices of an age or region, these poems are outright personal attacks.[96] In "Portrait van G. van D...... N," Van Rooijen portrays Van Doeveren as a poser who will quickly be unmasked:

> For some time he proudly parades with his cloak and Asclepius' rod,
> O joy! The robe falls from his shoulders
> There stands Aesop's crow with all his figments[97]
>
> ANONYMOUS, *"Portrait van G. van D...... N"*

The cloak and the Asclepius rod obviously are the traditional accessories of the academic physician, but the rod might also be explicitly mentioned because of the ambiguity of the symbol of the snake: an icon of both rejuvenation and poison, the bite of the snake resembles the potentially fatal sting of inoculation.[98] The mention of Aesop's crow refers to the fable of the fox and the crow: a fox sees a crow with a piece of cheese or meat in its beak, and flatters the crow into singing for him so it lets go of the food.[99] Despite the classical references, this short poem with its meek ABBA-ABBA-CCD-EEF scheme was

---

94 Daniëls, *De Kinderpok-Inenting in Nederland.*

95 Ibid., 99–108. Daniëls was given a set of document by Mr. W.P. Sautyn Kluit, whose grandmother was a granddaughter of Adriaan van Rooijen. The poems were anonymous, but Daniëls compared the handwriting to letters written and signed by Van Rooijen, and came to the conclusion that he must have been the author. He describes this on the first two pages.

96 Geerars, "Theorie Van De Satire."

97 Quoted in Daniëls, *De Kinderpok-Inenting in Nederland*, 99: "Hij stapt een wijl met 't pack, en AESCULAAP zijn staf / Trots voort; o vreugd! Het kleed glijdt van zijn schouders af / Daar staat Aesopus' kraay met all' zijn herssenprullen."

98 On the ambiguity of the snake on the Asclepius rod, see Ronson, *The New Medicine*, 122–32.

99 Aesop, *Aesop's Fables*, 112, www.forgottenbooks.org.

a far cry from the erudite medical-heroic Horatian odes of which Van Rooijen was certainly also capable.[100] Possibly he resorted to this crudeness to safeguard himself from suspicions. If he did so, it had the intended effect, because the adversaries of inoculation still had the upper hand in Leiden. This became clear on 1 May 1773, when the city council of Leiden once more decided to issue a written decree that banned the practice.[101] However, Van Doeveren persisted in advising colleagues in other cities on inoculation.[102] He did not respond with venomous poetry himself however, instead, like-minded friends stood up for him.

In 1777, six years after the start of the Leiden inoculation controversy and shortly after his royal inoculation success, a Latin poem dedicated to Van Doeveren and his inoculation practices was published by his friend Petrus Burman secundus (1713–1778), a classics scholar and neo-Latin poet who was a professor in rhetoric and national history at the University of Amsterdam, and who wrote and republished much classical Latin poetry.[103] That the inoculation debate was far from over is clear from the fact that this poem was immediately replied to with another poem by Cornelis Peereboom, an Amsterdam physician who was a strong adversary of inoculation. However, Peereboom's poem was once again in Dutch, and not as eloquent as Burman's by a long way. Burman used the classical poetic form of an elegiac distich, a Horatian-style lamentation in two-line stanzas of which one line is a hexameter, the other a pentameter.[104] In content, Burman effortlessly mixed Horatian characters and events with current affairs, as appears from this fragment:

Utque diu exemplo Byzantia turba praeivit,	And once the Byzantine troupes have taken the lead with their example,
Auxilia & Rheno Bosporis unda tulit,	And the water of the Bosporus has brought help to the Rhine,
Exprobrat haec Belgis, Othomana a gente doceri,	These Belgians are reproached for having let them be taught by the Ottoman people,

100 Van Rooijen published various volumes of Elegiac poetry in Latin during his life, i.e. *Poëmata* (1778).

101 Daniëls, *De Kinderpok-Inenting in Nederland*, 115.

102 I.e. Van Doeveren, MS UvA OTM: hs. Y123: 0.

103 Burman, MS SA 1776.

104 Luck, "Love Elegy," 405.

Quae simul ars miseris vulnus opemque ferat.	This practice that brings injury and help to the miserable ones at once.
Sic quondam Haemonia Pelidae saucius hasta	Like once wounded by the Haemonic spear of Achilles,
Telephus hac ipsa cuspide sanus erat.[105]	Telephus was cured by that same spearhead.

PETRUS BURMAN, "*Salutifera Variolarum Insitio*," 1776

Obviously, the first two stanzas refer to the transfer of the inoculation practice from the Ottoman empire to the Low Countries in the early eighteenth century: from the Bosporus to the Rhine. The last stanza, in contrast, refers to a classic legend that also appears in Horatius' Odes. According to this legend, Telephus, a son of Heracles, was wounded by Achilles' spear (Pelides is a patronymic for Achilles, son of Peleus) in the battle of Troy. An oracle informed him that the lethal wound could only be cured by what had inflicted it, and in exchange for directions to occupy Troy, Telephus persuaded Achilles to use his spear to cure him.[106] Obviously this myth was thought by Burman to be an almost perfect allegory for the prevention of smallpox by deliberate inoculation with the disease.

But why did Van Doeveren not respond in poetry himself? After an article on smallpox in the proceedings of the Hollandse Maatschappij der Wetenschappen in 1770, he appears not to have published anything further on the subject. And this at a time when it was not uncommon for medical professors to write both technical prose and lofty poetry aimed at achieving beauty on the same topic.[107] He expressed his frustration with the course of events in Leiden in letters to his friends, writing to Tak at one point: "I am ashamed of our Nation, often as I think about the unfortunate history of inoculation."[108] However, he seems to have made no attempts to publicly defend himself in poetry, although the Latin poem on friendship he wrote on leaving the Goes gymnasium in 1747 shows he would have been very eloquently capable of doing so.[109] Reasons for this reluctance may have been arrogance, but more likely it

105 Burman, "Salutifera Variolarum Insitio," 4.

106 Anthon, *A Classical Dictionary*, 1296, Nisbet and Hubbard, *A Commentary on Horace*, 171.

107 I.e. Van Doeveren's contemporary Gadso Coopmans (1746–1810) published both an inaugural speech on smallpox medication (1774) and a poem on the causes and control of the same disease, which was read to mark his resignation (1783).

108 Van Doeveren, MS UvA OTM: hs. Y123: 0, recto 1.

109 Van Doeveren et al. 1747.

was fear of further complicating his position as an inoculation advocate in a city where the practice was forbidden most of the time, and of not wanting to strain relations with Leiden colleagues like Van Rooijen anymore. In addition, in a 1776 letter to Salomon de Monchy, a Rotterdam physician who advised the city on inoculation, Van Doeveren expresses his view that too avid and careless lobbying for smallpox inoculation would only strengthen resistance against the practice.[110]

What remains curious about this episode is that whereas Van Doeveren based his advocacy of inoculation on favourable results from previous inoculations, his opponents seem to have screamed blue murder about inoculation without citing any circumstantial evidence regarding its effects. Apparently Van Rooijen was somehow less affected by aesthesis, particularly by the aspect of gaining knowledge based on one's own sensory perceptions. So how can Van Rooijen's aversion to inoculation be understood? The explanation is probably twofold. Van Rooijen was a religious conservative, and a pragmatic when it came to empiricism. This is clear, for example, from the introduction he wrote to a translated work on botany in 1745. He opens with the statement that a detailed knowledge of botany is essential for theologians, and only on the third page does he hesitatingly say something about the role of personal sensory perceptions.[111] Although Van Rooijen had to admit that seeing with one's own eyes and feeling with one's own hands was probably the best method to 'read the Book of Nature,' he found the idea that this was a feasible enterprise naïve:

> It is true that the one who works from experience lays the best foundation to build on with certainty, and can give good reasons for his actions, yet how much time passes before one, secured through one's own and repeated experiments, is really experienced. A lifetime is truly too short to check everything properly and to see the desired fruits of the labour applied.[112]

---

110 Van Doeveren MS UvA Y 123: l, Van der Zwaag, "Wouter van Doeveren" p. 142.

111 Van Rooijen, "Voorrede," i: "Want slaet men eens het oog op de Godtgeleerdtheidt welke de heilige blaederen doorzaeidt met gelykenissen hier van genomen ontvouwt wat grove misdagen begaet ze niet uyt enkele onkunde van Boomen Kruyden en veelvuldige andere Gewassen zo dat men wel zeggen mag dat een naeuwkeurige kennisse van deeze wetenschap voor alle uytleggers der gewyde Schriften niet alleen nuttig maer zelfs ten hoogsten noodig is..."

112 Ibid., iii: "T is waer dat die naer ondervindingen te werk gaet den besten grond legt om daer op met zeekerheidt te bouwen en goede reden van zyn handelwyze geeven kan maer welk een tydt verloopt er niet eer men hier in alleen door eyge en herhaelde proefneem-

Judging by Van Rooijen's attacks on Van Doeveren's pro-inoculation campaign and these statements, he must have felt that inoculation was an interference with God's ways, and the reliance on previous positive results of inoculation premature.

From this episode it appears Van Doeveren used aesthesis in the inoculation controversy in two ways. First, he remained true to his idea that knowledge should be primarily derived from one's own sensory perceptions and experiments—he himself had acquired good results with smallpox inoculation while in Groningen, so he felt confident advocating it once in Leiden. Van Doeveren thus stuck to his principles of basing his medical beliefs on his own sensory perceptions, the perceptions of others, and the use of reason by analogy, applied with the required cautiousness, instead of simply giving into the pressure of higher-ranking officials and colleagues. This also appears from the discussions he held with his colleagues in other cities, like Amsterdam, and from the extensive notes and tables he kept on his inoculations.[113] Second, he refused to give up good taste and manners even when he was bombarded with poetical squibs. The one poem published in his defense by his friend Burman was a paragon of poetic virtue, a true classics-informed discourse. This poem demonstrated the more learned and refined qualities of the pro-inoculation camp as opposed to those of the religious conservatives who lunged at Van Doeveren with quasi-satirical personal attacks in crude Dutch rhyme.

The ugliness of smallpox marks and the venomous verse of colleagues were not the only unpleasant things Van Doeveren had to deal with in his career. In fact, his is the first of the eighteenth-century Leiden anatomical collections that contained a substantial number of pathologies and monsters.[114] As we will see, aesthesis still defined Van Doeveren's way of working, and studying disease and deformation did not exclude searching for elegance and perfection, using sensory perception and materiality to understand anatomy. Because of the relative scarcity of deformed and diseased specimens, commodification

---

ingen beveiligt regt ervaeren wordt De leeftydt waerlyk is te kort om naer behooren alles na te gaen en ter belooning van den aengewenden arbeidt de wensehte vrugten te zien."

113 Van Doeveren, MS UvA UB: HSS-mag.: IV A 9: 2.

114 Until the 1770s, the Leiden anatomical collections only contained some pathologies and monstra that stemmed mainly from the seventeenth-century collections. The dominant Albinus collection hardly contained anything else other than normal anatomy, as Albinus' research interests had been focused on the ideal human body, its smallest structures, and the physiology of the senses. For more on Albinus' research interest see chapter 4. The LUMC catalogue lists a number of other historical monstrous preparations (Al0158–Al0183), but their origins and age are unclear. They are not listed in the catalogues of Anatomical Theatre, nor in the *Museum Anatomicum*.

also played an important part in collecting them. In a curious and unexpected way, monstrous preparations are the material remains of an aesthesis ex negativo, a search for perfection in the seemingly ugly and horrible.

The exploration of the materiality of a smallpox-marked ear and a monstrous little dog in this chapter has clarified several points regarding aesthesis and the ugly in eighteenth-century Leiden anatomy. First, the monstrous dog provides us with clues about the complicated utilitarian relation between humans and animals in eighteenth-century medicine and society. More importantly, from the way Van Doeveren dealt with monsters it appears he tacitly aimed to change the disgusting monsters and severed, decaying body parts into tasteless, odorless preparations through aesthesis. He used his own observations and sensory perceptions in as many cases as possible, both human and animal, to seek beauty, purpose, harmony, and regularity in what had long been viewed as freakish incidents or even the devil's work. In presenting the monsters as plain, natural, purposeful, inoffensive objects, Van Doeveren implicitly showed that they were not so much ugly, horrible, or disgusting but rather representatives of thus-far unknown categories. If anything, they were at most representations of a purposeful ugliness and thus had the same attraction as the sublime's little sister: the sublate. Because of their rarity and monstrosity, they bring home general truths in a particularly vivid manner. Their suggestion of death and decay, like that of all anatomical preparations, although tasteless and odorless, ensures that the insights they provide are grasped not only with the mind but also with palpable somatic resonance.

Second, it has shown that some diseases, such as smallpox, were so omnipresent in the eighteenth century that they had severe social impact and led to intense academic and social controversy. Ruysch thought it was necessary to cover up severe effects of smallpox, and not even Albinus, the greatest perfectionist of all, could avoid smallpox marks in his preparations. Behind this tiny smallpox mark lays the story of the huge impact smallpox had on both everyday and academic life in the eighteenth century, and the social and academic impact of new cures and practices. The intense controversy that arose between the fatalist adversaries and the supporters of inoculation, and thus human intervention, was, after all, a sign of changing ideas about death, disease, and the role of medicine. This chapter shows how interest in the deformed and the ugly refined the eighteenth-century understanding of aesthesis, and how this was reflected in controversy. Academics such as Van Doeveren and his supporters combined the refined discourse of classic Latin poetry and their reliance on their own sensory perceptions and observations to defend their position against adversaries less attached to the principles of observation and experiment, and to increase their standing as men of taste and learning. It would be superficial to distinguish between academic and poetic controversy here, as the publications all belong to the discourse surrounding one and the same controversy.

By the late eighteenth century, the making of anatomical preparations was still an act of selection, accentuation, and seeking beauty, even in the deformed and the ugly. By preserving monstrosities like these in preparations, and averting the immediate danger implied by the visceral disgust such specimens would provoke unpreserved, their makers shaped them into didactic instruments and purveyors of meaning. It could be said that these preparations are a visual and material variant of literary attempts by the likes of Goethe and Bilderdijk to make the damaged, decaying body an object of learning and aesthetic experience. Finally, this chapter has shown that Ad0022 and Ab0100 are material remains of eighteenth-century controversies, social issues, morality, and changing ideas about disgust, disease, and deformities. Only by answering the questions evoked by their materiality, can we fully understand how and why they were collected, created, and used.

CHAPTER 6

Colonial Bodies: Collecting the Exotic Other[1]

One of the most iconic preparations from the eighteenth-century Leiden anatomical collections is a male foetus of about five months old, wearing strings of black-blue and white beads around its neck, waist, wrists, and ankles.[2] It is now considered a 'top piece' from the historical collection, on permanent loan to the Leiden Museum Boerhaave. It has been assigned to the collection of Leiden anatomy professor Sebald Justinus Brugmans (1763–1819), and according to the catalogue description is 'African,' although its origins are unclear. And this beaded baby is not alone: another nine similar preparations of human foetuses decorated with beads are housed in Leiden, and four more are found in the Utrecht University Museum, the Amsterdam Museum Vrolik, and the St. Petersburg Kunstkamera respectively.[3] Apart from this preparation, several of the other foetuses have been documented in books on collections and collecting. One beaded baby has been pictured in three books in the past twenty years, in one case together with the very similar preparation from the St. Petersburg collection.[4] It is hard not to linger on these preparations, as they instantly evoke many questions: where are they from, and how did they end up

---

1 For a more general discussion on the fate of these and other colonial anatomical preparations, see Hendriksen, "The Fate of the Beaded Babies."

2 LUMC catalogue number Af0044. Malik, *The Meaning of Race*: Race is obviously a tricky concept, rather a social construction than a biological fact and now largely obsolete in the life sciences. Yet in the period that the preparations discussed in this chapter were made, 'race' was an only an emerging concept, first applied as a metaphor in matters of organic reproduction and used both for biological and biopolitical ends. See Mueller-Wille and Rheinberger, "Heredity," and Mazzolini, "Las Castas." Moreover, as B. Ricardo Brown, *Until Darwin*, 10, points out: "... the ability of science to fix—however unstable and temporary this might be—the classification of human variety has contributed mightily to the establishment of the authority of the science of life in our understanding of the truth about human nature and society." Terms like 'black' or 'African' and 'Asian' were used by eighteenth-century natural philosophers and natural historians in attempts to create categories within the continuum of variations of human appearances. As the preparations discussed in this chapter were defined in these terms by their initial collectors, I will maintain these categories here.

3 For the preparation in the Kunstkamera Collection see http://www.kunstkamera.ru/kunst-catalogue/index.seam?c=RUYSH , accession number MAE RAS: № 4070–817.

4 LUMC catalogue number Al0038, Kunstkamera Collection MAE RAS: № 4070–817. Fuchs and Fuchs, *Conserving*, Gould and Purcell, *Finders, Keepers: Eight Collectors*, Kemp and Wallace, *Spectacular Bodies.*

 | DOI 10.1163/9789004262775_007

ILLUSTRATION 6.0 *LUMC Af0044. Sebald Justinus Brugmans (1763–1819). Male foetus of about five months old, wearing strings of black-blue and white beads around its neck, waist, wrists and ankles. On display at Museum Boerhaave. Photo: author.*

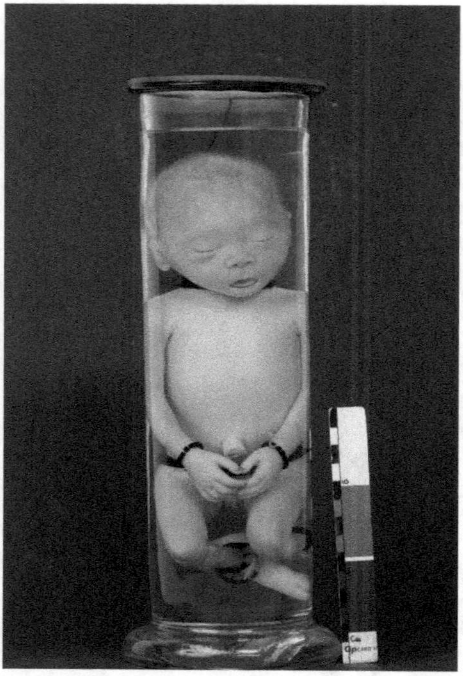

ILLUSTRATION 6.1 *Human foetus decorated with trading beads, from the collection of Gerardus Vrolik (1775–1859). © Anatomisch Museum Universiteit van Amsterdam.*

in these collections? Were they made here or abroad, were they made to order or unrequested gifts? Did aesthesis play a part in their creation, and why are preparations like these not found in other, similar collections from the same period? Why do they appeal to us so much?

In this chapter, an attempt will be made to answer these questions, and where reliable answers prove to be impossible, plausible possibilities will be suggested. Although it has proved impossible to trace the exact origin of the beaded babies, this chapter will show that there is much more to them than meets the eye, and that the aesthesis that dominated Leiden anatomy also influenced their creation and acquisition. Using the materiality of these preparations as a starting point will at least partially answer the pressing questions they evoke, and provide them with a historical context that is far richer than the few lines of text currently available. First, the preparations will be put in a theoretical and historical context: the appeal they have to us, the role of aesthesis and commodification in the creation and acquisition of these kinds of objects, and the few clues we have on their origins from written sources.

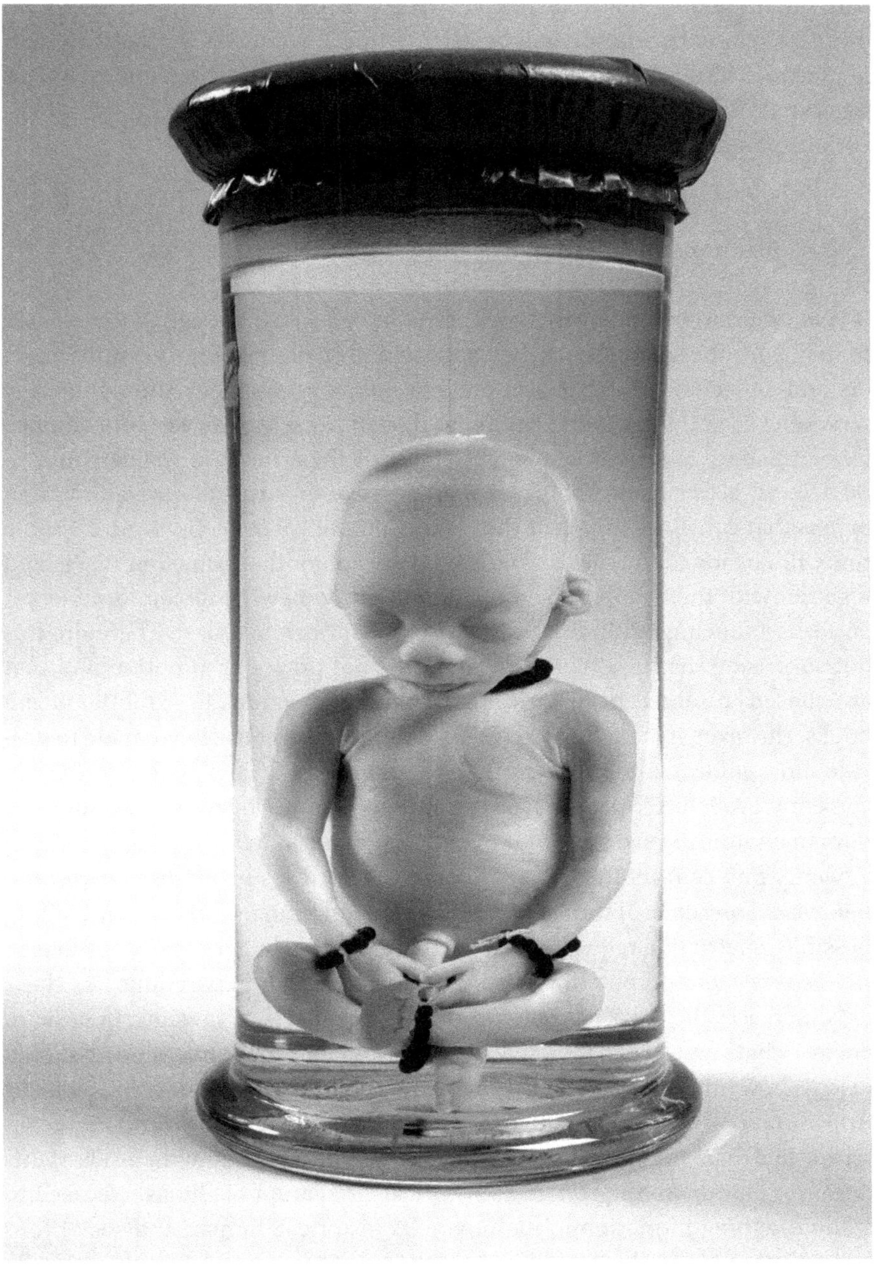

ILLUSTRATION 6.2 *Human foetus decorated with trading beads, from the collection of Jan Bleuland (1756–1838). © Universiteitsmuseum Utrecht, Photography: HetFotoatelier, Utrecht.*

Subsequently, the babies and their beads will be described, and the leads from that close description will be used to trace external evidence as to their origins and histories. Based on this, possible makers and circulations will be suggested. This will finally enable a reinterpretation of the meaning of these curious preparations.

Aesthesis and Commodification

It is not hard to imagine why these preparations are so appealing to us—it is probably for the same reason they appealed to their eighteenth-century makers and collectors. These babies are evocative because they simultaneously represent us and the other. They remind us of ourselves, as we were all once human babies, and their wholeness, (most of them have no visible congenital defects) is much less abhorrent than the severed, often diseased limbs and organs that constitute much of the rest of anatomical collections. At the same time, the beads the babies are wearing refer to another time and place, and together with their supposed ethnic origins are somewhat uncomfortable, yet curiosity-inducing reminders of a not-so-distant colonial past. Therefore it is not surprising that it is these preparations that draw our attention and that are chosen to represent historical anatomical collections in exhibitions and books. However, it is slightly surprising that no attempt has been made to provide more information on them than the fairly scarce catalogue information currently available. This is especially curious as most historical museums of human anatomy strive hard to prevent their objects becoming ingredients of a twenty-first-century freak-show. Even if the silence surrounding the prominent visual presence of these preparations is born from fear of controversy and possible repatriation requests, it is unwise to let them linger silently any longer.

Keeping silent on potentially painful, controversial preparations like these only increases the risk that they will end up as subjects of gawping. In order to create debate and transparency, researchers and museum professionals should discuss how museums should deal with and care for this kind of collection.[5] As Appadurai pointed out, although it is formally true that things have no meanings other than those that humans and their actions endow them with, this does not illuminate the concrete, historical circulation of things. We need to study the thing-in-motion to illuminate its historical, human and social context.[6] These objects do not talk; yet neither are they mute. What they tell the

5 Van Duuren et al., *Physical Anthropology Reconsidered*, 8.

6 Appadurai, *The Social Live of Things*, 5.

average visitor is more like a rumour than a story now, more confusing than clarifying. Tracing their story will amplify what they have to tell us to an audible level. It is fairly easy to identify the end products of knowledge making, but the active production of knowledge—however exciting—is more difficult to trace.[7] Moreover, as Roberts, Schaffer and Dear have shown in *The Mindful Hand*, the creation of both knowledge and objects are intimately related, sometimes inseparable, processes.[8]

The materiality of the beaded babies is the key to understanding their making and acquisition, as well as the knowledge they incorporate. Moreover, the creation and collecting of these foetuses was not simply rendering them stable, mobile, and combinable in order to ship them to the dominant centre of knowledge—as Latour would have it—but could also be an inquiry into and an acquisition of indigenous knowledge.[9] Yet first the role of aesthesis in the commodification and exchange of naturalia in the eighteenth century needs to be clarified, and some light must be shed on the unclear origins of these preparations.

Aesthesis in Commodification and Exchange

How did the beaded babies end up in Leiden, and in Utrecht, Amsterdam and St. Petersburg? On a conceptual level, the answer to the question of how the beaded babies ended up in the Leiden anatomical collections must be sought in the aesthesis aspects of commodification and exchange. Commodification of the dead human body—turning it into, or treating it as a commodity that is not by nature commercial—in natural history collections in Western Europe first occurred in the seventeenth century. This is clear from the presence of human bones and dried human organs in both institutional collections such as that of Leiden University, and in private collections such as that of Ole Worm.[10] In the case of the beaded babies, the fairly recently developed technique of making wet anatomical preparations for the first time allowed learned men to effectively commodify the dead body of the exotic other, making it possible to exchange it as a material fact.[11] Moreover, decorating these bodies with beads

7 Smith and Schmidt (eds), *Making Knowledge*, 3.

8 Roberts et al., *The Mindful Hand*, xvi.

9 Latour, *Science in Action*, 223.

10 See i.e. *Catalogues* 1669–1753 and Worm's *Museum Wormianum*, 1655.

11 Blankaart, *Anatomia Reformata*, 1695, 749 states that Longinus already advocated the shipping of human bodies and body parts form different regions, and that keeping them

and other accessories associated with their supposed geographical origins was a method for stabilizing them within the continuum of variations of human complexions.

The commodification of exotic human bodies was very significant for the development of the understanding of 'the other.' Whereas many accounts of indigenous people up to the mid-eighteenth century describe them in terms of barbarism and wild beasts, the natural philosophers who gained access to these dead bodies by the late eighteenth century were for the first time able to compare them to other human bodies. They immediately classified them as human, and started asking serious questions about the anatomy and ways of living of other people. Although it is very easy to classify these early explorations of physical variations, ethnicity, and other ways of living as racist, backward, and dangerous, we should not forget that they also provided the basis for modern anthropology. Commodification alone was not enough to give the learned men of Europe access to these object-bodies. In order to reach Europe, they had to be exchanged, something which had been happening with exotic naturalia such as insects, plants and animals ever since colonial explorations began in the late fifteenth century.[12]

Although the Dutch colonial trade was in severe decline by the middle of the eighteenth century, there was still a considerable stream of naturalia flowing to the Netherlands, sometimes upon explicit request, as appears from numerous accounts from the period. The minutes of the senators of Leiden University for example, show that in 1751, 1752 and 1754, the university received gifts of stuffed birds and insects from the African governors Swellengrehel and Rijk Tulbagh.[13] From a 1766 letter by the latter to Arnout Vosmaer, the custodian of the Royal cabinet of naturalia, it appears that Tulbagh sent the Stadholder

---

covered in a coffin filled with lye or brine of sea salt and alum would ensure they arrived at their destination without corruption or decolouration. Yet these bodies seemed to have been intended for dissection, and probably were not suitable for ongoing preservation. This is further supported by the fact that Frederik Ruysch (1638–1731) appears to have been the first to successfully create lasting wet preparations, as has been discussed in chapters 2 and 3, but he does not explicitly list wet preparations of bead-decorated black foetuses. He only listed one foetus of unspecified origins decorated with exotic plants and animals (*Thesaurus Anat.* VI, 41–2, no. LXI) and one apparently undecorated black foetus (*Thesaurus Anat.* IX, 21, no. XXXII). In the Ruysch collections in St. Petersburg, an embalmed foetus with the remains of a flower wreath around its head remains, but the origins are unclear (MAE RAS 4070-832).

12 For commodification and exchange of naturalia in the Dutch Republic in the seventeenth century, see Cook, *Matters of Exchange*.

13 Suringar, "Het Bijeenbrengen Eener Verzameling," 288.

boxes of insects and shells, and on one occasion even a live pig.[14] Not only royalty and universities received such exotic gifts, as shows from the fact that VOC physician and directing member of the Batavian Society Jacob van der Steege (1746–1811), a former student of Petrus Camper, sent his old teacher a skeleton, four skulls, a tongue, and a penis of a Javanese rhinoceros from Batavia.[15]

Neither was the exchange of objects between colleagues in different European countries exceptional, as appears, for example, from a letter by Andreas Bonn. In 1804, he wrote to his friend Altheer that he had heard the latter had the intention to presently send some books to Johann Friedrich Blumenbach, who was the first to categorize humankind into five races. If possible, Bonn wrote, he would like to add a small chest to that mailing, which "would not be heavy and which would contain a number of dry pieces for Professor Blumenbach's Cabinet."[16] None of the sources cited explicitly mention the shipping of wet preparations of human anatomy, yet it is very possible these were exchanged as well—after all, they somehow ended up in Leiden by the early nineteenth century. Although the commodification and exchange of naturalia, including (parts of) human bodies, was nothing new, I will now argue that aesthesis did play a particular role in the commodification and exchange of the Leiden beaded babies.

As Cook pointed out, ever since the early seventeenth century Dutch intellectuals aimed to employ reason combined with the five senses when describing and explaining nature, but the objective facts that were the result of that method could also take on attributes of taste and discernment.[17] This combination is strongly present in aesthesis, where knowledge is gained through sensory perceptions combined with reason, and where the investigator simultaneously and largely tacitly searches for beauty while also trying to eliminate the ugly and the disgusting. In the case of the beaded babies, this shows in the fact that they are, as wet anatomical preparations, products of investigation and preservation, which were equally informed by the senses, reason and a quest for beauty.

They are informed by the senses in so far as creating such a preparation involves all of them: the tactility of obtaining and handling a foetus, looking at how it is formed and how it sits in its phial, hearing the sounds of making the

---

14 Tulbagh to Vosmaer 1766, Leiden University Special Collections BPL 246.

15 Snelders, "Het Bataviaasch Genootschap," 82, Rookmaaker and Visser, "Petrus Camper's Study."

16 Bonn MS KB 1804, recto: "Het kistjen soude niet zeer zwaar zijn en eenige droge stukken voor het Cabinet van den Professor bevatten."

17 Cook, *Matters of Exchange*, 410.

preparation, smelling and maybe even tasting the preservation liquid for the right consistency. They are informed by reason, as the maker must have theorized about how old the foetus was, why the beads were already on it or had to be put on it. And finally, his actions were informed by his quest for beauty and perfection in considering how he would have to make the preparation look and how to preserve it in order to make it an attractive commodity; not as something appalling, but as an object which would gain him something—be it a nice piece for his own collection, a tradable object, or an appreciated gift for a colleague or friend. This leads us to ask when and where were these aesthesis-influenced commodities created and exchanged, whether there were more of them, and why so many eventually did end up in the Leiden anatomical collections.

Unclear Origins

Surely there were more eighteenth-century anatomical collections that contained foetuses of black or 'African' children. For example, Frederik Ruysch already listed three foetuses of 'American Negroes' in his collections.[18] The 1799 sales catalogue of the Middelburg city physician Egbert Philip van Visvliet (1763–1799) mentions "a fully grown negro child of the female sex" and "ditto of the female sex, of five or six months," and a 1869 catalogue of the Zeeuwsch Genootschap der Wetenschappen (Zealand Society of Sciences, est. 1769) lists a "Foetus of a negro."[19] In the 1779 collection catalogue of the German professor Wagler, a friend of the Leiden anatomy professor Wouter van Doeveren, an "Embryo abortus Aethiopis" is listed.[20] In Petrus Camper's collection at Groningen University we also find a "negro foetus."[21] As is the case for these collections, the information on the beaded babies in the modern Leiden University Anatomical Museum is rather scarce. The first descriptions of the preparations in the Leiden Anatomical Museum catalogues are meagre, yet at least they give some kind of indication as to when these preparations first

---

18 Ruysch, *Alle werken*, 1389, 1447.

19 Visvliet, "Catalogus Eener Uitmuntende Verzameling," 175: "1. Een volwassen Negerkind van de Vrouwelyke Sexe. 2.—dito van de Vrouwelyke Sexe, van vyf à zes maanden." De Man, "Naamlijst," no. 150, Zeeuws Archief 26.1.02.2.162, suggests that this preparation was donated by D. Van Cruysselbergen, a Leiden graduate, in 1778.

20 Wagler, *Verzeichnisz Der Anatomischen Präparate*, 8, no. 104.

21 E0605, Cabinet 3 in the Camper collection at Groningen University: http://anatomie.ub.rug.nl/index.htm.

appeared in Leiden and from which private collections they originated. Five of the preparations have been traced back to Eduard Sandifort's 1793 *Museum Anatomicum I* and son Gerard Sandifort's 1835 *Museum Anatomicum III* respectively.[22] This was done by curator A.M. Luyendijk-Elshout in the 1940s and 50s, who analysed the original labels of the flasks and phials where possible.[23] One of the preparations listed in the *Museum Anatomicum III* is probably "A Foetus of a female negro, purchased from sir Voltelen in 1795," mentioned in an inventory of the anatomical theatre from 1798.[24]

Eduard Sandifort first describes Ad0030 under the 'Supellex Anatomica Doeveriana' on page 108 of *Museum Anatomicum* I. He mentions nothing about the small string of coloured beads around the neck of the foetus, but simply describes it as "African foetus of six months. Colour not black, but darker than European."[25] The apparent confusion about the skin colour of the preparation can simply have been caused by the discolouring effects of the preservation liquid. As this was the only 'African' embryo in the collections there was also little to compare it to. In the Albinus collection there were some preparations of black skin of adults, but these were mostly injected, whereas the embryo was not, making comparison difficult.[26] Forty years later, the collections of the Amsterdam professor Andreas Bonn (1738–1817) and the Leiden professor Sebald Justinus Brugmans, which had been acquired by Leiden University in 1822 and 1819 respectively, were described in *Museum Anatomicum III* by Sandifort's son Gerard. Suddenly, there are many more preparations of black foetuses and black skin, and in the Bonn collection they even appear to constitute a separate category.

In the Brugmans collection on pages 27 and 28 preparations Af0044 and Af0103 are listed. The first is described as "Similar African fetus of five months," 'similar' referring to a previously listed preparation which is described as "male foetus of four months, very fleshy."[27] So once again, there is no mention of

22 Sandifort, *Museum Anatomicum*, vol. I: Ad0030, MA III: Af044, Ac0100, Af0103 and Ac0113.

23 Elshout, *Het Leidse kabinet*. All preparations are contained in lead glass phials, a type that was in use from the late seventeenth century until the middle of the nineteenth century.

24 A.C. 228, printed in Witkam, "Over De Anatomieplaats," 63. This is probably Floris Jacobus Voltelen (1753–1795), the son of a minister in Cape Town who was ordinary professor of medicine and chemistry at Leiden University from 1783 till 1795.

25 Sandifort, *Museum Anatomicum*, vol. I, 108: "CLXIV. Foetus aethiopis semestris. Color non niger, Europaeo tamen obscurior."

26 Sandifort, *Museum Anatomicum*, vol. I, 27–90.

27 G. Sandifort, *Museum Anatomicum*, vol. III, 27: "CCCIX. Foetus similis aethiopis quinque mensium." ('similis' refers to: "CCCVI. Foetus masculinis quadrimestris, torosus admodum.)

beads. On the next page, Af0103 is described as "African foetus of the same shores with part of the umbilical cord and placenta."[28] The umbilical cord and placenta have probably been removed due to conservation reasons over time. In the section of normal (naturali) wet preparations, four Bonn preparations of 'African' babies are listed under three pieces of black skin on page 331 of *Museum Anatomicum* III. Sandifort junior described them as follows:

> XIV. African foetus of about seven months; a few curly hairs; the colour of the skin is black; compressed nose, lips are swollen.
> XV. African foetus of about six months, excellently representing the character of the nation; the colour is not very black.
> XVI. Foetus of the same age as the female Hottentot; colour of the entire body very black.
> XVII. African foetus of about seven months; colour of the skin is less black, but offers an example of several of the African peoples; the back is very hairy.[29]

Remarkably, all these preparations have survived until the present day, but only the first is decorated with beads.[30] Together with 'the Hottentot' it is currently on display in the Anatomy Museum in the historical cabinet, whereas the other two preparations have been put in storage.[31] Finally, there is one beaded baby in the Bonn collection that is also currently in storage, and which differs from the others in three respects. This is a dried preparation of a foetus with anencephaly and spina bifida, listed by Sandifort under the dried

---

28 Ibid., 28: "CCCX. Foetus aethiopis ejusdem actatis cum sune(?) umbilicali et placenta."

29 Ibid., 331: "XIV. Foetus Aethiopis septem fere mensium; capilli parum sunt crispi; color cutis fuscencit; nasus est compressus; labia sunt tumida. XV. Foetus sex mensium Aethiopis, characterem nationis egregie exhibens; color non valde fuscus est. XVI. Foetus ejusdem aetatis Hottentotae; color totius corporis fuscescit. XVII. Foetus septimestris Aethiopis; color cutis minus fuscus, sed caeteras characteres Aethiopum offerens; dorsum valde est pilosum."

30 XIV, now registered as LUMC catalogue number Ac0100.

31 XVI is current LUMC catalogue number Ac0015. G. Sandifort, *Museum Anatomicum*, vol. III, 331, preparation XV has record number Ac0014 in the current catalogue; XVII has number Ac0016. A. van Dam, curator of the Anatomical Museum, has suggested that the 'very dark' skin colour of the 'Hottentot' preparation might in fact gave been caused by mercury or silver salts added to the preservation liquid. This is worth investigating in the future.

pathology preparations.[32] It has traces of dissection on the stomach, and the only beads on it are placed on the eyes.[33]

Then there are the five remaining beaded babies that have never been assigned to a specific sub-collection.[34] According to Luyendijk-Elshout, three of these were also listed in the manuscript catalogue made by Teunis Zaaijer (1837–1902) in the 1860s, but this catalogue was partly lost and only a summary per cabinet from 1892 remains.[35] Zaaijer himself had a keen interest in the emerging discipline of anthropological anatomy. He gained his MD in Leiden on 30 May 1862 with a thesis on two female pelvises from the East-Indian Archipelago.[36] In 1893 he made a manuscript catalogue of the 'Racial skulls, pelvises and skeletons in the Leiden University Anatomical Cabinet.' In the preface to the listings of preparations, he warns against 'falsification' of anthropological specimens: he gives the example of a custodian who marked the skull of a man born in Leiden as 'Russian' so he could fetch a higher price for it.[37] This example shows that the commodification and exploitation of supposedly real 'exotic' dead human body parts continued well into the nineteenth century.

To make the tracing of the origins of these preparations even more complicated, no written sources are known—books, pamphlets, letters or otherwise—by Leiden anatomists or their contacts which mention them. However, the fact that preparations Al0040 and Al0041 cannot be traced to the *Museum Anatomicum* catalogues, and were not found in the Zaaijer manuscript catalogue by Luyendijk-Elshout, does not necessarily mean that they are indeed 'false' or younger than the other beaded baby preparations. That can only be made clear through an object-driven analysis of the preparations themselves.

---

32 LUMC catalogue number Ac0113.

33 G. Sandifort, *Museum Anatomicum*, vol. III, 399: "DCLXXXII. Foetus aliquot mensium acephalus cum tota spina bifida, Balsamo conditus atque exsiccatus est."

34 LUMC catalogue numbers Al0037, Al0038, Al0041, Al0045 and Al0040.

35 LUMC catalogue numbers Al0037, Al0038, Al0045. LUMC catalogue, 2011, Luyendijk-Elshout, *Het Leidse kabinet*, MS AM Boddaert 1892 only mentions: "Zaal 12—Kast 1. Rasfoetus enz. ttt1-ttt61 61 Spirituspraeparaten (foetus van negers enz. huid van negers, haar van negers en inboorlingen uit den Oost-Indischen Archipel)."

36 Zaaijer, *Beschrijving van twee vrouwenbekkens*, Thomas, "Teunis Zaaijer," 1499.

37 Zaaijer MS AM 1893, p. VII.

The Babies and Their Beads

The one thing that connects these ten babies is the beads and their supposed origins, but they are all in a way different from each other. The odd one out is the only dried preparation, and the only preparation with a visible birth defect.[38] The others are all wet preparations, preserved in 80% ethanol, and have no visible birth defects. In the eighteenth- and early nineteenth-century Leiden anatomical collections, they were mostly defined as 'black,' 'African' or 'Hottentot,' and 'Asian,' although their physical ethnic features are very difficult, if not impossible, to discern. These preparations were very relevant for contemporary research. Scholars like B.S. Albinus started wondering and seriously investigating what could explain the apparent variation in skin colours and physiognomy among people, and that these variations seemed to follow geographical patterns. Therefore it is justified to use the words 'African' and 'Asian' to describe these preparations here, as their original keepers saw them as exemplifiers of these categories.

Of the nine wet preparations, seven appear to be of African descent and two have Asian features or have been described by Leiden curators as such—one of the 'Asian' preparations has even been described in the mid-twentieth century catalogue as 'Mongolian,' which in Blumenbach's original racial classification system was the denotation for the 'yellow' race; we would probably say Asian.[39] Of the seven African babies, two are male and three are female, and of two it is hard to tell the sex as their legs are folded over their genitalia. Although some of them look very similar, they are all of different ages, varying between a four-month foetus and a neonate. Neither are any of the bead strings the babies are wearing exactly identical in colouring and location on the body. What, then, can be deduced from the way these babies look, from the beads they are wearing, and the way they were preserved? In the following, the preparations will be discussed grouped by their suspected origins, as that may provide the most feasible leads to their heritage.

The African Babies

The biggest 'group' of beaded babies in the Leiden anatomical collection are the supposedly African babies. Both male and female, from a four-month old foetus to a neonate, are wearing strings of beads around their neck, wrists,

38 LUMC catalogue number Ac0013.

39 Elshout, *Het Leidse kabinet*.

elbows, waist, loins, knees, and/or ankles. However, the colours and locations of the beads vary widely. The fact that the preparations are wet and still well-preserved means that they must have been made after the beginning of the eighteenth century, as the first who was able to make this kind of long-lasting wet preparations was Frederik Ruysch, who shared his secrets only with a select number of his confidants. Although Ruysch certainly came across black foetuses in his Amsterdam practice, it is unlikely any of these preparations are his, as he injected all his preparations—a method which applies to none of these.[40] By the early to mid-eighteenth century, more people started to learn how to make a lasting wet preparation, for example some Albinus' students. As most preparations were first mentioned in a Leiden catalogue somewhere between 1793 and 1860, they must have been made roughly somewhere between 1750 and 1850. So how did seven black babies wearing strings of beads of various colours on various body parts end up as anatomical preparations in a Leiden collection somewhere between the mid-eighteenth and early nineteenth century?

The most imprecise answer possible, based on the scarce information available, would be: they were preserved and presented or sold by someone with sufficient knowledge of making wet anatomical preparations, who also had access to aborted, miscarried, or still-born black babies (or babies that could pass as black) who were either decorated with strings of coloured beads by their bereaved mothers, or who were decorated with those beads by the person who preserved them or some reason. Who, then, would have access to those babies? Obviously, during the period in question, the second Dutch West Indian Company (1675–1792) had the monopoly on trade at the West-African Gold Coast, shipping slaves to the Caribbean, gold from Africa and sugar from the New World back to Europe.[41] Many of these slaves died either in the 'depots' where they were kept before and after shipping, or during the voyage.[42] Although officially forbidden, assault and rape were not uncommon aboard the slave ships, and even when pregnancies were carried full term and children born alive, their chances of survival on the ships were minimal.[43] Infanticide

40 Ruysch, *Alle Werken*, 726: To my knowledge, Ruysch did not list any uninjected, bead decorated, 'African' foetuses in his collections—only the 'head of the child of a negro woman', affected with small pox, which he decorated with injected arteries to make it less repulsive.

41 Den Heijer, *Naar De Koning Van Dahomey*.

42 Ibid.

43 Priester, *De Nederlandse Houding*, 107–114, Eijgenraam, "'Menschlievendheid En Eigen Belang'," 91.

followed by suicide was also not uncommon among the slaves, as appears from WIC archives.[44]

Moreover, Londa Schiebinger's 2004 book on colonial bioprospecting in the Atlantic world has convincingly shown that there is evidence that "women in the colonies, especially slave women and women of colour, practiced abortion (...) though at what rate and for what reasons is less clear."[45] Any European involved, therefore, in the slave trade would likely have had some kind of access to dead foetuses and newborns of slaves. Moreover, although small numbers of black people ended up in Europe, the beads with which these preparations are decorated are so-called trading beads, used as currency in the African trade, which suggests the preparations would have been made on the African west coast. The person able to do so was most likely a European medical man stationed in one of the trade centres, appointed by the Company to attend to the newly arrived slaves, whose health was, to a certain extent, of importance to their future owners.

Although these likelihoods narrow down the possible scene of action a little, this still leaves us with quite a number of putative preparation makers. Before starting to trace personal files and correspondence of all doctors and surgeons employed by the West Indian Company between 1750 and 1792, it might therefore be useful to see if the beads give any further indications of where the babies might be from. As mentioned before, the beads are so-called trading beads, used as currency in the African trade for centuries. The seventeenth-century centre of bead-production was Venice, but by the eighteenth century the beads were produced in many places, including the Dutch cities of Middelburg (until 1760), Veere, and Vlissingen.[46] As Trivellato has pointed out, establishing the origins of glass beads has therefore proven to be an arduous, if not impossible task.[47] Beads like these would be widely available at Dutch trading posts, but would also be in frequent use with African people all over the continent by the late eighteenth century, giving no clues about what particular tribe would use beads in this way.[48]

In addition, variation in beadwork between ethnic groups in Africa is endless. Even within one group there are different styles and colour codes. The meaning and fashion of bead colours vary over time, but have only been

44 MS NA: *Stukken Gediend Hebbende Bij De Criminele Processen Op De Kust Van Guinee. Afschriften.* 1.05.01.02 151–2. Tweede West-Indische Compagnie (WIC).

45 Schiebinger, *Plants and Empire*, 142.

46 http://www.geschiedeniszeeland.nl/themas/slavernij/handelsgoederen/kralen?lng=nl.

47 Trivellato, "Venetian Glass Beads," 65.

48 Van der Zwan, *Oog Voor Kralen*, 18. Dubin, *The Worldwide History of Beads*, 119–22.

ILLUSTRATION 6.3 *A pair of Female ere ibeji twin figures in the permanent collection of The Children's Museum of Indianapolis.*
Courtesy of The Children's Museum of Indianapolis. Image cc by-sa 3.0 The Children's Museum of Indianapolis.

mapped sporadically.[49] However, the bead decorations on the Leiden babies strongly resemble those of so-called ere ibeji, or twin statuettes, both in colours and location on the body. The bead strings around the necks, waists, wrists and ankles of these statues immediately remind us of the beaded babies in Leiden. The Yoruba people of West Africa have made statues like these at least since the late eighteenth century.[50] Whereas in the seventeenth century the Yoruba reportedly killed or abandoned twins and their mothers as they were considered the offspring of adultery and bad luck, by the late eighteenth century this had completely reversed and turned into a cult of twin-worship.[51]

The Yoruba have an exceptionally high incidence of twin births: one in every twenty-two births is a twin birth, whereas this rate is only one in eighty in most western countries. Since the eighteenth century, twins are considered the bearers of fertility and good luck, and are believed to have exceptional

49 Carey, "Gender in African Beadwork," 88–9.

50 Stoll and Stoll, *Ibeji. Zwillingsfiguren Der Yoruba*, 42–3.

51 Dapper, *Naukeurige Beschrijvinge*, 488: "Daer worden gene twelingen gevonden, doch, gelijk gelofelijk is, wel geboren; want het vermoeden is dat een van beide by de vrou-moer vermoort wort, aengezien het baren van twee kinderen t'eener drachte een vrouwe daer te lande tot grote schande strekt, alzo zy vastelyk geloven, dat een man geen vader van beide kinderen kan zyn." Also see Vogelzang, *Ere Ibeji. Beelden Van Tweelingen*, 5–6.

powers that they can also use against those they dislike. Therefore, it is of utmost importance to keep twins happy, with beautiful clothing, the best food, and social privileges. Twins are also believed to be sharing one soul between the two of them, so if one dies without having been able to lead a full life, it is very important that his or her soul is offered a shelter in which it can house and lead that life afterwards, so the soul of the remaining twin will not experience imbalance. This shelter is the ere ibeji, a person-shaped statue carved out of wood of maximum 30 cm (10 inches) height. It is carried around and nurtured either by the mother or the other twin if it is a girl old enough to do so, as if it were the infant itself. When both twins die in infancy, ere ibeji are made for both of them and the mother nurtures them and carries them with her everywhere, which also increases her social status. The ere ibeji are more often than not decorated with strings of beads around the neck, waist, ankles, and knees, and are sometimes given a string of cowry shells, which used to be used as currency, as a sign of wealth.[52]

The strings of coloured beads worn by the ere ibeji are not merely decorative; the colours have particular meanings and functions. Red and white beads honour Shango, god of thunder and lightning, blue ones Oshun, the goddess of life-bringing water and healing. White beads refer to Obatala, the creator-god, green ones to Orunmilla, the god of the oracle, a combination of black and red beads is to honour Eshu, deity of fortune, death, and travellers. A variety of colour combinations means the maker wants to honour and unify multiple deities. Especially noteworthy is that black beads around the waist are a protection against Abiku, restless spirits that are believed to take over the bodies of children who die before or quickly after birth, only to be born to the same parents over and over again.[53]

The Dutch stationed in West Africa in the seventeenth and eighteenth centuries may very well have seen bead decorated ere ibeji: the Yoruba then lived in an area called Dahomey and Oyo, roughly in what are now Nigeria and Benin. It is known that this area was occasionally visited by the Dutch slave traders to do business and solve conflicts with the local chiefs, for example as revealed in the letters and diary of WIC emissary Jacobus Elet from the 1730s.[54] Interestingly, a number of the tribes living in this area at the time appear to have been cannibalistic and to have used human skulls and bones for decorations.[55] However, no accounts of WIC personnel obtaining any of

52 Stoll and Stoll, *Ibeji. Zwillingsfiguren Der Yoruba*, 36–47, 70–71.

53 Al0037, Vogelzang, *Ere Ibeji. Beelden Van Tweelingen*, 18.

54 Den Heijer, *Naar De Koning Van Dahomey*.

55 Schlothauer, "Kopfjagd Und Schädelkult," 119–22. In fact, human body parts are a sought-after commodity even today in this area. See Ololade, "Where Body Parts," Aransiola & Asindemade, "Understanding Cybercrime," 762.

these decorations or other human material have been found. Another possible point of access to black foetuses was of course the slave ships. For example, a 1788–9 slave ship surgeon's journal of the Middelburg Commercial Company, a smaller trading company operating next to the VOC and WIC, lists numerous births, of which a remarkable number are multiple births.[56]

Reliable eighteenth-century descriptions of the people and customs of Africa are few and far between though. Very often, such accounts were compiled from the writings of others by so-called fireplace travellers, endlessly recounted and copied without any verification.[57] However, there is one written source by someone who at least visited the area himself, suggesting there was a fairly common tradition of decorating new-borns with beads on the African west coast in the eighteenth century. In 1709, Willem Bosman, chief merchant at the fortress Elmina (currently Ghana), issued a *Precise description of the Guianese Gold- Ivory- and Slave Coast* in Dutch. The book would be translated into English and French, and frequent reprints appeared well into the 1730s, making the book one of the more trusted and widespread sources on Africa from the first half of the eighteenth century. In the book, Bosman notes that:

> Once a child is born, the priest or offering-papist, called 'Fetischeer' or 'Consoe' here, is called for, who immediately orders the child to be dressed with strings, corals and other rags around the body, neck, arms and legs, which are sacred and blessed with his normal statements, which protect the child against all diseases and other discomforts.[58]

---

56 i.e. in MS ZA MCC 1410, the 1788–89 journal of the ship's surgeon of the Middelburg Commercial Company ship ZeeMercuur 1, of the nine births described two are multiple births. We find an entry on the birth of triplets of whom two are alive ("Een Vrouw Slaaf in de Kraam bevallen van drie jongedogters twee leevende een dood") on 29 May 1788. By June 14th, the two other girls had died too. The birth of a male twin is registered on 22 June 1789, one with a cleft palate and blind, dies straight away, his brother two days later ("Een Vrouw: in de Kraam bevallen van twee jongesoons het eene met een hasenmond en all .bij oogen blind op daatoo overleeden"). Three of the six single births were still births, and eventually all the babies born aboard the ship died within weeks.

57 Adams, *Travelers and Travel Liars*.

58 Bosman, *Nauwkeurige Beschrijving*, 118: "Zo haast een *Kind* geboren is, werd den *Priester*, of *Offer-Paap*, alhier *Fetischeer*, of *Consoe* genaamt, ontboden, dewelke aanstonds last geeft, dat men het *Kind* een gehele dragt met *Touwetjes, Couraal* en andere vodderyen om het *lijf, hals, armen* en *beenen* zal binden, 't geen dan door hem met zijn gewoonelijke *Beweeringe* werd *gewijd*, en *beswooren*, waar mede het als dan tegens alle *ziekte* of andere *ongemakken* is gewapend." (hyphens Bosman)

This description of the function of bead decorations on newborns is remarkable, as WIC personnel was generally distrustful of native medicine and religion—something which is still reflected here in the use of the demeaning terms 'offering-papist' (Bosman was a Protestant) and 'Fetischeer,' derived from the Portuguese 'feitiço,' magic.[59] The fact that Bosman associates the bead decorations applied to newborns with protection against disease and discomfort implies that this habit would have been of particular interest to western medical practitioners exploring the unknown.[60] Moreover, these examples strongly suggest that the bead decorations on the Leiden preparations are either original or inspired by the bead decorations of ere ibeji or West African babies. The accounts explaining why the maker of these preparations decided to decorate them in this particular way may have been already out-dated at the time of their creation, a fact of which he may also have been aware. Yet the decorations were important for the commodification aspect of aesthesis. Preserving the body itself ensured it became stable and tradable, and adding bead decorations perceived as 'African' in Western Europe was effectively a claim about the geographical origins of the babies, a way to purify and identify them as 'African.' Without the bead decorations, it would be very hard to discern anything particularly African about them, but with the beads, the babies fitted nicely within the new categories coined by Blumenbach and his contemporaries.

The Asian Babies

Something similar goes for the 'Asian' babies, although their decorations are even more layered. Preparations Al0045 and Al0040 from the Leiden anatomical collections are two female foetuses of respectively five-and-a-half and seven months old, with Asian features. One of them, Al0040 holds a twig with a berry in its right hand. The manner in which they are decorated with strings of beads is similar to the decorations on the African foetuses mentioned above, though this does not tell much about their ancestry. As noted before, the use of trade beads in itself (also called seed or pound beads) says little about who made the bead decorations on these preparations, or about where and when they were made. From the sixteenth century onwards, trading beads also played an

59 For the attitudes of WIC personnel on indigenous medicine and customs, see Rutten, *Dutch Transatlantic Medicine*, 101, continued.

60 Recent work by Snelders, *Vrijbuiters van de heelkunde*, illustrates that European medical men had a profound interest in materia medica and healing practices in the colonies, such as the use of talismans.

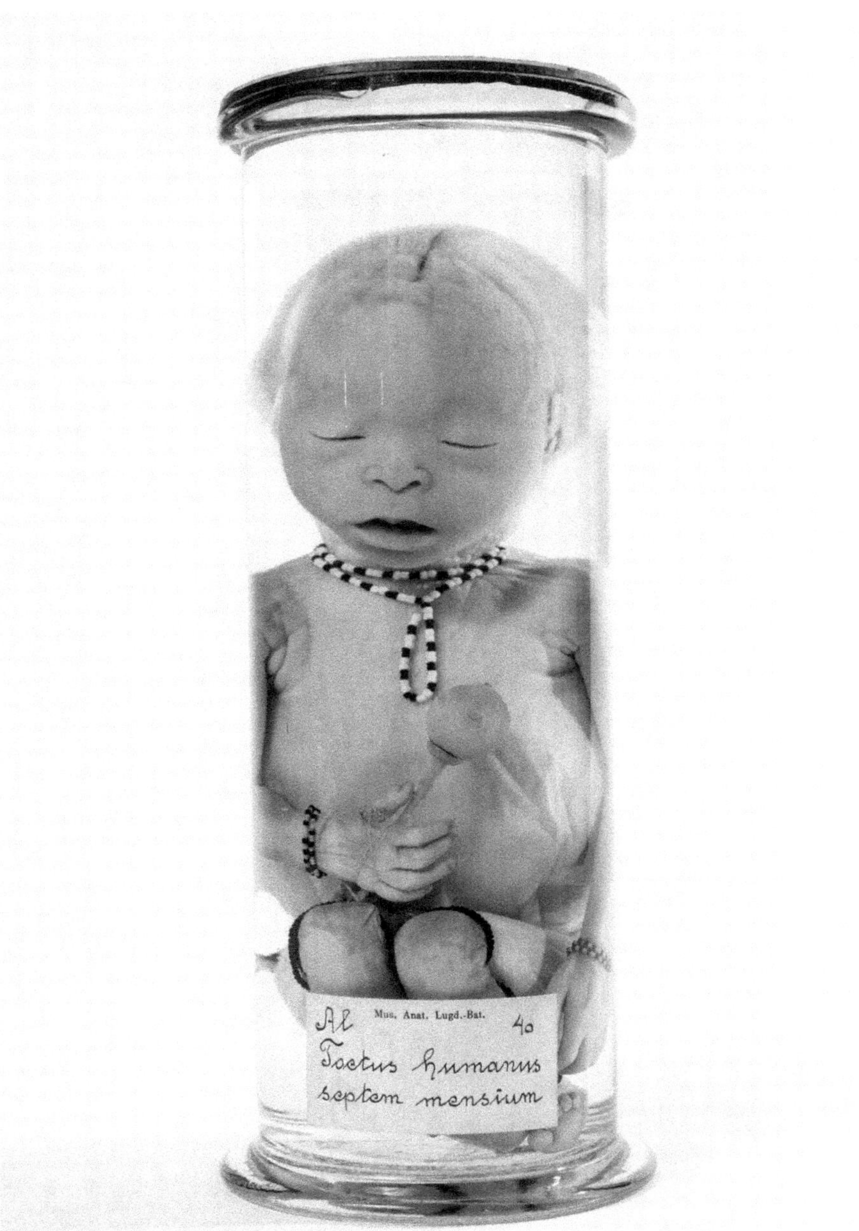

ILLUSTRATION 6.4 *LUMC Al0040. Female foetus of approximately seven months old, white, yellow and black beads around wrists, loins, knees, ankles, twig with berry (probably nutmeg) in right hand, held like a rattle. Probably from the collection of Sebald Justinus Brugmans (1763–1819). © Anatomisch Museum LUMC 2012. Photographer: Arno Massee.*

ILLUSTRATION 6.5 *LUMC Al0045. Female foetus, circa five months old. Wearing multicoloured beads and a yellow beaded necklace with a coin marked 'Hollandia 1778'. Probably from the collection of Sebald Justinus Brugmans (1763–1819). © Anatomisch Museum LUMC 2012. Photographer: Arno Massee.*

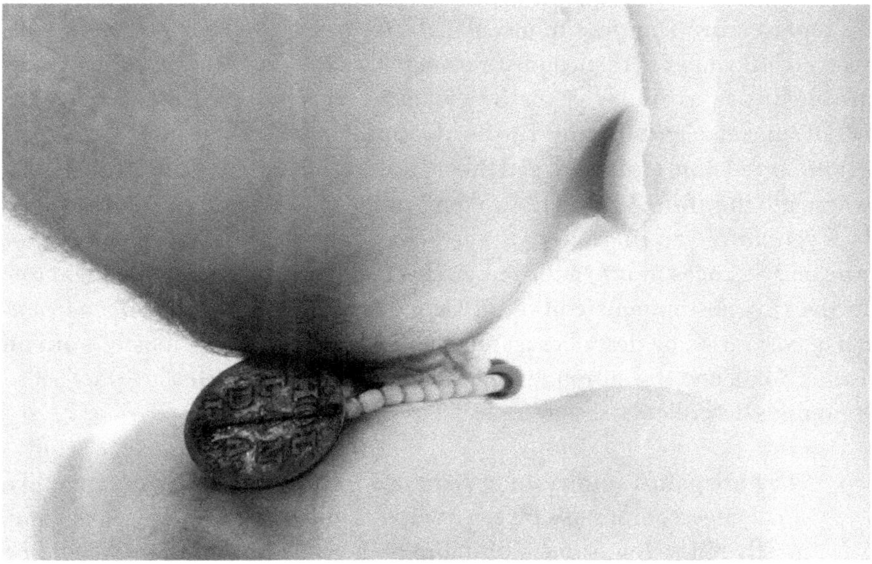

ILLUSTRATION 6.6 *Detail of Al0045 (Illustration 6.5): coin marked 'Hollandia 1778' on necklace. Probably from the collection of Sebald Justinus Brugmans (1763–1819). © Anatomisch Museum LUMC 2012. Photographer: Arno Massee.*

important part in trade and crafts in the East Indies, and beadwork would play an important role in rituals and dress in the area for centuries to come.[61] It is therefore possible that the decorations were either made by natives from the East Indies, or that they were made by a medical officer of the Dutch East India Company (VOC) who was familiar with similar decorations from the African West, coast or who had seen or even made the preparations of African foetuses before. After all, the only eighteenth-century route from Western Europe to the East Indies was one with stages of stops in Africa. Moreover, there is one particularity of the bead decorations of Al0045 that confirms the idea that this preparation is from the East Indies, or at least represents the area.

In a lengthy article from 1899, Gerrit Pieter Rouffaer (1860–1928), a versatile Dutch scholar, discusses the origins of "the mysterious moutisalahs (aggri-beads) in the Timor-group." At the time he published his piece, Rouffaer wrote, both men and women wore certain kinds of beads of various colours, though preferably yellow or orange, as jewellery. West Timor had been Dutch territory since the Dutch drove out the Portuguese in 1640, with a short British interregnum between 1812 and 1816. Could the yellow strings of beads on Al0045

61 Dubin, *The Worldwide History of Beads*, 238.

be representing Timorese moutisalahs? The beads Rouffaer describes have pierced little holes in them and are strung "like coral," and the resulting strings are rare treasures of the rich on these islands—that is, West Timor and a number of smaller islands nearby. The beads appear to be antique; they are no longer imported from elsewhere and the material is not indigenous to Timor.[62] So where are they from? The earliest mention Rouffaer identified are two pieces by a certain W. van Hogendorp in the Proceedings of the Batavian Society for Arts and Sciences in 1779 and 1780. An even older source mentions them too. In the 1705 posthumous edition of Georg Everhard Rumphius' *Ambon cabinet of curiosities*, he describes a strange kind of stone that is "mostly worn on Timor, Solor, and the surrounding islands."[63] The exact origins are unclear, so Rumphius describes how they look:

> They are grains of diverse appearance, most like slim disks, some like small cheese, other like pieces of pipes, similar to white or yellow coral, mostly red-yellow, some a bit more pink, some lighter, so they could be confused with worn corals if they would have had that shape (…) striped on the outside and pipe-shaped.[64]

That these beads were a desirable commodity appears from the fact that they were sold at half their weight in gold, and two strings of them would buy you a slave. As it was thought that the material for the beads could be found on the island of Sawo, west-southwest of Timor, the VOC tried to acquire the mining rights to it, but lost a ship in the attempt and subsequently abandoned the idea. No curative powers appeared to be associated with the beads, and the exact location where they could be found remained unclear. Rumphius grumpily notes that it was possible the Timorese simply refused to disclose this, as "it is a very angry Nation, that also hides other valuable crops and metals of their Land from us."[65]

Subsequently, the next description of the yellow beads seems to be the one by Hogendorp in the late eighteenth century, quoted by Rouffaer. Willem van Hogendorp (1735–1784) was a co-founder of the Batavian Society. He had obtained a law degree in Leiden in 1754, and in 1772 left for Batavia to make

62 Rouffaer, "De Raadselachtige Moetisalah's," 409.

63 Rumphius, *D'amboinsche Rariteitkamer*, 245.

64 Ibid., 246–7. On the next page he mentions that the name Mutu Labatta or Mutu Labatte is Malay for 'stone pearl', and only used by the Dutch and the 'Indians', the Timorese and Rottinese called them Va-Boa and Tzeda.

65 Ibid., 247.

up for the losses of the family fortune. After doing so, the ship he travelled back on, Harmonia, sank with all hands in 1784. In the same year, his two-part description of the island Timor appeared in the Transactions of the Batavian Society. In it Hogendorp wrote that "Each kingdom has its exceptional valuables, that consist of gold and silver plates and stringed corals" and that the indigenous population of the isle of Savo, near Timor, rubbed themselves with fragrant wood oils, especially nutmeg oil, and that the women wore girdles of black or yellow beads.[66] Rouffaer also quoted numerous sources from the nineteenth century. A certain Salomon Müller in 1829 wrote:

> Apart from the arm-rings and gold plates, the necklaces are the most highly esteemed decorations of the Timorese. They consist of small gold balls and a special kind of beads of a yellow-orange colour, hard like glass, unevenly rounded and pierced with little holes. Superficially seen, one would think it is sea coral, but acid has no effect on them.[67]

The descriptions of, and speculations on, the nature of the mysterious yellow-orange beads of the Timorese keep popping up throughout the nineteenth century, until the Berlin Professor Virchow in 1884 finally (and most likely correctly) identified the beads as carnelian.[68] This semi-precious stone is indeed not indigenous to Timor, but it is to India, so it is not at all unlikely carnelian beads ended up in Timor long before the arrival of the Portuguese in the early sixteenth century. The maker of Al0045 may very well have aimed to represent the yellow-orange moutisalah beads of the West-Timorese with the strings of yellow trading beads, combined with some other colours. Moreover, there is another remarkable aspect of the decorations of Al0045 that suggest a connection with the late eighteenth-century East-Indian Archipelago under Dutch rule.

Most remarkable is that Al0045 wears a beaded necklace which has been closed with a coin that has an inscription which reads "Hollandia 1778." (See illustration 6.6.) The coin on the necklace is by far the most vocal part of these preparations: it can be quite easily identified as a VOC double wapenstuiver

66 Van Hogendorp, "Vervolg Der Beschryving," 68: *"Elk Ryk heeft ook zyne byzondere kostbaarheden, die in gouden en zilveren plaaten, en aan elkanderen gereegen koraalsteenen bestaan." / p. 95: "Zy bestryken zich het lighaam met oliën van welriekend hout, en vooral van den noot-muscaad, waar van zy veel werks maaken* (…). *De vrouwen draagen eenen gordel, om het lyf, van zwarte of geele koraalen, twaalf of veertien vaemen lang."*

67 Rouffaer, "De Raadselachtige Moetisalah's," 412–13.

68 Rouffaer, "De Raadselachtige Moetisalah's," 426–30.

(weapon penny), coined in 1778 in Holland to be used exclusively in the East Indies.[69] These coins were produced by the thousands and being valuable nor rare, (they are nowadays quite widely available and sell for about ten euros), the coin may have been put on the necklace much later than 1778. In fact, it may have been put on the necklace at a later date with very specific intentions. By the late eighteenth century, the fourth English war (1780–1784) and its effects increasingly disturbed the VOC business. Shipments of money from the Netherlands did not reach Asia in time, or not at all, and return shipments with goods did not reach the homeland. In the latter days of the VOC, from 1796, the Company was therefore issuing emergency money. These 'whoppers' ('bonken') were stamped square pieces of copper, chopped-off Japanese bars of the metal.[70] The VOC penny as decoration on the necklace of the foetus could thus well have been a visual reference to the glory days of the VOC.

Moreover, a close look at Al0045 reveals that whereas the coin looks as a lock at the back of the necklace now, it was probably originally at the front, as a sort of medallion. A greyish denture on the foetus' chest certainly suggests so. This is remarkable in the context of a report by a former secretary of Timor and its subordinates. H.D. Wiggers in 1892 sent the Batavian Society of Arts and Sciences a string of moutisalahs with an account of how these were worn on Timor. In it, he wrote:

> The most expensive kind is that of dark yellow or orange colour and of regular size. Overall, the moutisalah is light yellow in colour. Said beads are also worn by married women as a sign of unavailability, with a silver tobacco box attached to it as a medallion. If the required number of beads for a necklace is not available, the strings are commonly completed with normal Chinese or European manufactured beads.[71]

The Tropenmuseum in Amsterdam indeed has a number of eighteenth-century round silver tobacco boxes from the East Indies in its collections which could very well be used on a necklace as a medallion. With this information, Al0045 suddenly appears to be not simply a specimen of foetal development in the 'Asian race,' but simultaneously an example of a Timorese woman in traditional jewellery and an explicit reference to colonial riches and power.

---

69 Cuhaj & Dudley (eds.), *Standard Catalog of World Coins*, 1119. Many thanks to Marcel van der Beek and Ans ter Woerds, coin curator and librarian at the now no lpnger existing Geldmuseum Utrecht for helping me identify this coin.

70 Pol, *Schepen Met Geld*, 46.

71 Rouffaer, "De Raadselachtige Moetisalah's," 443. (translation mine)

Then there is the twig with the berry held by Al0040, which resembles a rattle.[72] If this would have been a Ruysch or Albinus preparation, the ball-shape of the berry would probably have been an allegory for the vanitas homo bulla—man is nothing but a bubble. However, this preparation is likely to be at least half a century younger, and the meaning of the berry-rattle therefore must be sought in its colonial background and in general iconographic themes. Although it is hard to determine a berry that has been kept in preparation liquid for about two hundred years, this specimen appears to be nutmeg. As we have seen, Hogendorp mentioned that some natives perfumed themselves with nutmeg-oil.[73] Nutmeg also played an important role in the colonial trade in the East Indies. It was not only appreciated for its culinary characteristics, but also stocked on the slave ships for other uses during the journey. Nutmeg has active hallucinogenic properties, which the slaves knew could relieve the strains of the long journeys, and the Dutch themselves recommended it for all kinds of physical ailments. However, excessive use could cause delirium and even death.[74] Naturalists and medical men, both in the colonies and the Netherlands, were therefore very interested in nutmeg.[75] Apart from the possible medical importance, the iconography of the berry as a rattle may refer to the traditional function of a rattle given to infants to scare off evil spirits.[76] This would fit with the known meaning of the beaded strings as described in Bosman's travel accounts: in a way the African beads and the western 'rattle' have the same function.

Moreover, the common or fragrant nutmeg (*Myristica fragrans*) is native to the Banda Islands of Indonesia, which were brought under Dutch control with brute force in 1621. This meant the Dutch would dominate the nutmeg trade for almost two centuries, until the English temporarily gained control over the Banda Islands during the Napoleonic wars (1803–1815).[77] Nutmeg was highly valued in Europe, and the French too went to great lengths to get hold of and cultivate nutmeg plants.[78] As nutmeg was one of the spices traded in

---

72 Luyendijk-Elshout in 1952 identified this preparation as 'Mongolian', but her reasons for doing so are unclear.

73 Van Hogendorp, "Vervolg Der Beschryving," 95.

74 Rumphius, *Het Amboinsche Kruid-Boek*, vol. II, 21–3, Rutten, *Dutch Transatlantic Medicine*, 46, Cook, *Matters of Exchange*, 126.

75 On the role of Dutch medical men in collecting colonial naturalia and material medica from the seventeenth century onwards, see i.e. Swan, "Collecting Naturalia," Van Gelder, *Tussen Hof En Keizerskroon*, and Weber, *Hybrid Ambitions*.

76 De Vries and De Vries, "Rattle," 246.

77 Cook, *Matters of Exchange*, 186–7.

78 On these attempts see Spary, "Of Nutmegs and Botanists."

considerable quantities by both the WIC and the VOC, it was probably chosen by the maker of this preparation—like the weapon penny on the other baby—as a symbol of the (by then withering) glorious days of the Dutch East India Company. In fact, although it appears that the baby with the coin necklace was already in the Leiden collections by the 1860s whereas the lack of mention in the Zaaijer catalogues suggests the baby with the nutmeg was a later addition, there is a strong indication that these two preparations were actually made by the same person, maybe even as a pair.[79] The nutmeg 'rattle' has been tied crisscross to its hand with a piece of string; on the right hand of the other supposedly Asian preparation, we see an almost identical crisscross imprint of string too.[80] Although we will probably never know what was in that hand, it could well have been a sprig with a berry of the sandalwood tree, native to Timor. Another possibility is that this preparation also originally held a nutmeg 'rattle', referring to Rumphius' mention of a belief among the local women that nutmeg could help women conceive twins.[81]

As the bead decorations, the berry 'rattle' and the coin 'medallion' on the two Asian preparations are most likely to have been added by a European who was not only interested in human anatomy and anthropology, but who also wanted to create vivid reminders of a recent, more affluent, past. Once again, the commodification of the human body was not only a way to render it stable and tradable, by decorating it the maker also made claims about the geographical and ethnic origins of the body, and by shipping the thus stabilized and categorized body back to the dominant centre of knowledge he ensured he stayed connected to it. Because of the similarity in the bead decorations and the way they were preserved, and combined with their limited geographical spread nowadays, it seems likely that the Asian babies were preserved by the same person as the African ones; or at least by someone who was familiar with the latter. As mentioned previously, it appears that aesthesis, and particularly the commodification aspect associated with it, played a part in the decisions to decorate preparations of human foetuses this way. In fact, aesthesis turned out to be essential in the commodification and exchange of these preparations.

79 LUMC catalogue numbers Al0045 and Al0040 respectively.

80 LUMC catalogue numbers Al0040 and Al0045 respectively.

81 Rumphius, *Het Amboinsche Kruid-Boek*, vol. II, 15.

Possible Makers

The scarce eighteenth-century descriptions of West African and East Indian people and their customs means that the beads on these preparations tell us little, which brings us back to the question of how they ended up mainly in Leiden. Which European medical man or men could have brought or sent these preparations back to the Netherlands? It would be helpful to know more about these people, as this could also tell us more about the commodification aspect of aesthesis. In the correspondence of the Leiden professor of anatomy Sebald Justinus Brugmans (1763–1819), a number of contacts in Africa that provided him with skulls and 'curiosities' are documented, but these all date from after 1800.[82] Nor do any of the letters specify the nature of the curiosities, so these sources do not tell us anything about the beaded babies, and the only possibility that remains is to make deductions.

If these foetuses were indeed from Africa and Asia, where may they have come from; who would have been in a position to transport them to the Netherlands? For the African foetuses, it was most likely through someone stationed in one of the trade centres on the West African coast or the West Indies, appointed by the West Indian Company in the second half of the eighteenth century. Moreover, this person would have had access to the foetuses of African babies, he would have known collectors of anatomical preparations in the Netherlands, more specifically Leiden-educated anatomists in Amsterdam, Utrecht and Leiden, who would have valued such preparations. Also, this person would have had sufficient skill to lastingly preserve them. Trying to identify someone who fits this profile presents us with another problem: the meagre administration of the second WIC. By the second half of the eighteenth century, the trade on the West African coast had been opened up to all residents of the Dutch Republic; only the slave trade with Surinam, Essequibo, and Berbice was still exclusive to the WIC.[83]

Unlike the VOC, the WIC hardly kept track of who was employed on its ships and in its stations abroad. Surgeon-majors were likely to be experienced barber-surgeons; practically trained medical men with little or no academic education in medicine, well-versed in operations such as cutting stones and performing amputations. In general, it was hard to find qualified staff for the WIC, as the journeys to Africa and the West Indies were long and arduous. Often

82 Van Heiningen, *Correspondence*, numbers 207, 229, 259, 551. http://www.dwc.knaw.nl/wp-content/bestanden/brugmans.pdf.

83 Den Heijer, *Naar De Koning Van Dahomey*, 169–71.

the death rate among the staff was higher than that of the slaves they shipped.[84] At the VOC, it appears the positions of ship's surgeon and assistants were more popular, which might have had something to do with the reputation of the company and the possible financial incentives the VOC offered.[85] Between 1770 and 1790, the VOC employed over fifty ship's surgeons.[86] This leaves us with numerous candidates indeed, many of whom left little or no material traces which can confirm or exclude them as the initial collectors of these preparations. However, the person (or persons) who made these preparations was likely to have some kind of connection to Leiden University. Lacking comprehensive evidence for these preparations' ancestry, two suggestions of possible makers will be discussed here: Jacob Voegen van Engelen and Cornelis Terne.

Johan Voegen van Engelen (1755/6–1792?)

The son of a Baptist minister, Jacob Voegen van Engelen was enrolled at Leiden University in September 1771. His study of medicine in Leiden ended abruptly in the spring of 1776, when it appeared that his girlfriend Clara van Meurs, the daughter of one of the substitute bailiffs, was pregnant. As his father refused his permission for a wedding and made (eventually unsuccessful) attempts to have Johan deleted from the *Album studiosorum*, the couple fled to Den Bosch in the southern Netherlands. Jacob then went to Duisburg, where he obtained his MD degree within two weeks of his arrival, on 9 May 1777. By 1778 Jacob and Clara had apparently returned to Leiden, as their son Cornelis was buried there in a rented grave on 17 May. On 4 October 1778, the couple finally got married in the English church. In the next four years, they had three more children, none of whom reached maturity. Although Voegen van Engelen developed as a rather successful writer and publisher in the years to come, it seems that without the financial support of his father he had trouble maintaining himself and his family, as he took up jobs on the side, like assistant city physician and doctor of the Swiss students.[87]

From the few letters in his hand that remain, it appears the aging Petrus Camper, the Leiden educated professor of Anatomy based in Groningen, who was an old friend of his father, was a sort of substitute father and Maecenas to him. In his letter of 19 January 1781, Voegen van Engelen addresses Camper as 'mon cher Protecteur,' and merely eleven days later, he writes to him and tells

84 Ibid., 156.
85 Bruijn, "Ship's Surgeons," 206–7.
86 Ibid., 313.
87 Lindeboom, "Inleiding Bij De Surinaamsche Artz," X–XVI.

him how comforting Camper's care for their 'domestic happiness' is to him and his wife.[88] From the letters it becomes clear that Camper corrected his printing proofs and answered his questions on new publications in the field such as that by Monro on the nerves in the Philosophical Transactions of 1780. Voegen van Engelen, in his turn, took Camper's side in a controversy over the dissection of an orang-utan from the Royal Menagerie in 1779.[89] However, Camper's support was likely to have been mainly emotional and practical rather than financial. This probably explains why, around 1785, Voegen van Engelen eventually accepted the position of lecturer in the Surinam colony.

From Surinam, Voegen van Engelen published a journal on the influence of the tropical climate on health. Entitled 'De Surinaamsche Artz' (The Surinam MD), it appeared between 1786 and 1788, but appears to have not been read widely. This is also the last sign of life of Voegen van Engelen. Did he send bead-decorated foetuses back to his alma mater from Surinam? It seems unlikely given the manner in which he left Leiden, yet the undecorated black foetus in the Camper collection could well be a gift from him to his old friend and protector. If we take the distinguishing bead and plant decorations on the Leiden foetuses as an indication that they were all preserved by the same person, a person who wanted to position himself in the Leiden aesthesis tradition, we should look for someone who spent considerable time both in Africa and the East Indies, who had an interest in foetal development and the colonies and the exotic. Cornelis Terne is therefore a more likely candidate.

Cornelis Terne (1747–?)

Sieving through the WIC archives has revealed that the surgeon-majors (chirugyn-majoor) employed by the company, according to the 1792 minutes of the council of the West Indian Colony, were obliged to keep a certain set of instruments during their appointment, to which end they were given a two hundred guilder grant.[90] Together with the printed minutes that outline the instrumental requirements for the surgeon-majors in the WIC archive, I found two remarkable manuscript copies from the minutes of the council of the WIC from 22 February 1793. The first is a request to appoint a certain Cornelis Terne, 'Medicinae Doctor and man-midwife' as surgeon-major with the third

88 Voegen van Engelen 1781, MS UL BPL 247/26.

89 Voegen van Engelen 1779, MS UL BPL 885.

90 Extract found in: MS NA *Kamer Amsterdam. Register Van Uitgaande Brieven Aan Diverse Personen En Instanties Hier Te Lande. Met Bijlagen, Met Onvolledige Tafel En Onvolledig Repertorium.* 1.05.01.02 462. Tweede West-Indische Compagnie (WIC).

battalion of the Curaçao garrison. Because of his outstanding knowledge and experience, it is requested that he will be given an extra ten guilders a month during his stay in the colonies, an extra twenty-four guilders a month during his journey, (another indication of the harshness of life aboard), as well as an additional two hundred guilders one-off fee and a 'free act' of a hundred guilders. The second manuscript is a confirmation of said request, in which an even higher stipend is established, signed the next day, 23 February 1793.

These documents are remarkable because Cornelis Terne was indeed a very experienced and knowledgeable physician and man-midwife. In fact, up till that moment, Cornelis Terne had been city obstetrician (vroedmeester) in Leiden. Born in May 1747 in Amsterdam, he enlisted with the VOC in 1766, disembarked in Africa in 1768, and by 1770 had returned to Amsterdam to marry a certain Barendje van Buuren. Apparently she died soon thereafter, as in 1771 he married Sophia Christina Grave, with whom he had four children.[91] They moved to Leiden in 1776, where Terne obtained his MD with a thesis on abortions and premature births in 1778.[92] His supervisor was F.B. Albinus, and he dedicated his thesis to him and his other teachers: G. van Doeveren, E. Sandifort, J.D. Hahn, N.G. Oosterdyk and A. Bonn. In the same year, Terne was appointed as city obstetrician, a post he would hold for almost fifteen years. Apparently, Terne was not only very learned and experienced, he had a big ego and a temper too. In a 1990 article, Van der Borg has traced the many clashes he had with the city midwives and the city medical council, which eventually led to his dismissal in 1793.[93] In the same year, he divorced his wife, and, as it now appears, he enlisted with the WIC.

Are the African foetuses in the Leiden collection Terne's work? As there are no known records to confirm this, we will probably never know for sure. However, Cornelis Terne is a very likely candidate. First, as far as we know, he was the only medicinae doctor with a Leiden degree enrolled by the WIC in the second half of the eighteenth century. This means he probably had acquired the technical skills to create long-lasting wet anatomical preparations during his Leiden training, something unlikely for a barber-surgeon. Second, Terne's Leiden training makes it very likely he was familiar with the aesthesis of the Leiden anatomists who were his teachers and masters: Albinus and Van Doeveren. Also, he would have known other Leiden-educated anatomists like Brugmans, Bonn, and Bleuland. Terne may well have seen Ad0030 in the collection of his teacher Van Doeveren. This very light-skinned foetus has only a

91 For dates and archival pieces see Van der Borg, "Het Ontslag," except enlisting with VOC: http://vocopvarenden.nationaalarchief.nl/detail.aspx?ID=1270029.

92 Terne, *Antwoord Op De Vraag*.

93 Van der Borg, "Het Ontslag."

simple string of coloured trade beads around its neck, and may have inspired Terne years later to obtain more beautiful, more perfectly 'African' specimens of fetuses for his former teachers and their colleagues—after all, the beaded babies can first be traced to the collections of Brugmans and Bonn.[94]

The influence of aesthesis can be seen in all these preparations: they present material, tangible knowledge of 'the other' and 'the exotic' as well as knowledge of normal human anatomy, providing in a sensory manner perceivable objects which would otherwise be almost inaccessible for the eighteenth-century European anatomist. Also, the bead decorations of these apparently perfect babies fit well with the aspect of aesthesis which seeks to display beauty and elegance in the material products of one's knowledge, and thereby of averting the possible disgust dead bodies evoke. Whether Terne may have added the beads or they were already on the babies is not really relevant—either way, they made the babies into elegant, distinctly foreign preparations. Finally, Terne's original training as an obstetrician, with primary focus on fetal development and premature births, easily explains his possible interest in these African fetuses. After his enrollment with the WIC, only one more trace is found of Cornelis Terne: in 1794 he won a gold medal for his reply to a competition issued by the Batavian Science Society, on the feeding of infants in the tropics. He is still mentioned as 'obstetrician of Leiden' in this publication. Whether Terne indeed ended up in Batavia can only be revealed in Indonesian archives, but if he did, that might also explain the presence of Asian foetuses in the Leiden collections.

What remains a mystery is the one beaded baby in the St. Petersburg Kunstkamera. It has been assigned to the collection of Frederik Ruysch, but it is unlikely to have been a Ruysch preparation because of the lack of injection, and it is not mentioned in the original Rusych catalogues.[95] However, exchanges of knowledge between Russia and the Netherlands did not come to an end with the death of Peter the Great in 1725. For example, Abraham Kaau Boerhaave (1705–1758), a cousin of the famous Herman Boerhaave, was professor of anatomy and pathology at the St Petersburg academy between 1748 and 1745.[96] A few decades later, Leiden graduate and governor Johan Meerman (1753–1815) travelled extensively through northern Europe with his wife from 1779 until 1800. They spend two winters in St. Petersburg and

94 Sandifort, *Museum Anatomicum*, vol. 1, 27–90, 108.

95 Ruysch, *Alle werken*, does list three foetuses of 'American negroes,' but no decorations. As he does explicitely mention bead decorations ('paarlen') in other preparations, it is highly unlikely that the beaded baby in the Kunstkamera, or any of the other beaded babies, are originally from his collections.

96 Sjtsjedrova, Irina. "Abraham Kaau-Boerhaave."

Meerman described his visit to the Ruysch Collections in the Kunstkamera.[97] Although he mentions nothing about donating preparations to it, he was part of the extensive Leiden network and even had a cousin, L. de Sille, who was a ship's surgeon with the VOC.[98]

These kinds of travels and connections were certainly not unique in the late eighteenth century; another indication of the ways in which anatomical aesthesis could spread around the world. What unifies all these men—Abraham Kaau Boerhaave, Johan Voegen van Engelen, Cornelis Terne and Johan Meerman—is that, as Latour would say, they wanted to domesticate the minds of others and stay connected to the dominant centre of knowledge.[99]

In this chapter, a number of questions raised by ten eighteenth-century preparations of foetuses thought to be of African and Asian descent in the Leiden anatomical collections have been answered. First, the uncomfortable appeal of these preparations and the role commodification and exchange played in their creation were discussed. This made clear that commodification of the babies was an inquiry into indigenous knowledge as well as a way to stay connected to the dominant centre of knowledge. Moreover, it was a manner to fix otherwise fluent variations in human appearance into somewhat stable categories of geographically determined human race, an emerging field at the time.

Subsequently, the materiality of the preparations was used to establish that the real or desired geographical origins of these babies in the Leiden collections must be sought on the eighteenth-century West African coast and in the East Indies, and possibly also the West Indies. The hypothesis that they were collected, decorated, and preserved by a medical man or men employed by the West Indian Company and, in the case of the Asian babies, the East Indian Company (VOC) has been discussed. Although there is no definitive proof of it in the Netherlands, the former Leiden city obstetrician Cornelis Terne is a very likely candidate. The uniquely high concentration of these preparations in the Leiden Anatomical Museum collections, combined with their lush decorations of beads, a coin, and a plant, is a strong indicator that they were part of the aesthesis so characteristic of eighteenth-century Leiden anatomy. However, their decorations function only partially in the same way as the decorations in the Ruysch and Albinus preparations. They also commodify the preserved body and direct the gaze of the observer to some extent, but in these preparations they are also act in such a manner as to communicate political, cultural and

97 Meerman, *Eenige Berichten*, 204.

98 Bruijn, "Ship's Surgeons," 151, MS MMW De Sille 1802, 1804.

99 Latour, *Science in Action*, 215–20.

medical knowledge from the colonies to the centre. The occasional presence of similar individual preparations in other collections can easily be explained through the connections with Leiden networks.

The object-driven analysis in this chapter has shown three important things. First, that the materiality of the object itself is an extremely valuable source in tracing and understanding the active production of knowledge, especially when few other sources are available. Questioning the object itself shows once again that the creation of both knowledge and objects are intimately related, and mostly inseparable processes. The Leiden beaded babies do not simply represent knowledge; they are both the result of knowledge production and eighteenth-century knowledge about human anatomical variety itself. Second, the analysis in this chapter shows that even though we have no written accounts of the makers or first keepers of these objects, and thus cannot even be sure that they actually are 'African' or ' Asian,' they were certainly the product of aesthesis. They were equally informed by sensory perceptions, reason, and a quest for beauty and perfectiom. This 'colonial' aesthesis demanded the recording of the physiognomy and habits of largely unknown people to enable further analysis.[100] The recording itself—the making of the decorated preparations—was a very sensory and tactile affair, which in addition required a sense of perfection, as only a perfectly preserved, well-decorated specimen would make a worthwhile commodity.

Finally, the obscured meanings of the bead decorations on the African foetuses and the anthropological significance of the beads, plant, and coin on the Asian foetuses show that the preservation, decoration, and collecting of these foetuses was not simply done to render them stable, mobile, and combinable in order to ship them to the dominant centre of knowledge but was also an inquiry into, and an acquisition of, indigenous knowledge, and a way to categorize them. Objects like these beaded babies cannot be understood properly through meagre catalogue descriptions, nor do glossy pictures in coffee-table books do them justice. When studied closely and respectfully, they tell vivid histories of changing and emerging disciplines and ideas. However uncomfortable and confronting these histories may be, they do need to be told and heard, as they offer us invaluable insights into our past and lessons for the future.

100 I put colonial in parenthesis here because I do not want to pretend I have written a history of colonial aesthesis or anatomy. The babies are at best a tiny fraction of colonial aesthesis in anatomy—a topic that definitely deserves more scholarly attention.

CHAPTER 7

Back to the Bone: The End of Aesthesis

Aesthesis in eighteenth-century anatomy was characterized by several factors: gaining knowledge through sensory perception; searching for perfection and elegance; dealing with disgust by either using visual or literary strategies; seeking systems and meanings in the negatives of deformation and pathologies; and a stabilization and categorization of the human body through commodification and decoration. All these aspects can be retraced through the materiality of the eighteenth-century Leiden anatomical preparations. Aesthesis in anatomy was a particular epistemic culture, tacit and fluid. Like any epistemic culture, it did not last forever. It is as hard to discern when it first appeared as when it eventually disappeared. The end of aesthesis seems to have coincided approximately with the end of old anatomy and physiology as described by Cunningham.[1] In this chapter, the materiality of a very particular kind of anatomical preparations enables us to understand the end of aesthesis: bone preparations.

The preparations discussed so far were mostly wet preparations of soft tissue. There are several reasons for this. First, wet preparations form a substantial part of the collections; over thirty per cent of the eighteenth-century preparations remaining today are preserved in fluid.[2] Second, wet preparations were still a relative novelty at the start of the eighteenth century, and thus developed and changed much faster than the already existing dry preparations and preparation techniques, which had been around since the Egyptians first created mummies. Finally, wet preparations are often more visually appealing and evocative than dried preparations, as dried preparations of soft parts are often especially difficult to identify for those without medical training. Many wet preparations, with their lifelike looks of pinkish and red soft tissue, or their decoration of lace-rimmed sleeves and bead necklaces, are more easily recognizable as humans or parts of humans than dried soft tissue preparations, and are thus easier to identify with.

Although wet preparations look more clearly manufactured than dry preparations, the latter, either of soft tissue or bones, are still preparations, not some

1 Cunningham, "The Pen and the Sword."

2 2490 preparations of 8170 in total.

 | DOI 10.1163/9789004262775_008

kind of objets trouvées.[3] Bones constitute a particularly interesting ontological problem here, as they initially appear to be just that: bare bones. Yet in the process of preparing, using, and displaying bone preparations, their materiality changed; they became artifacts and were invested with new meanings. As Hallam has pointed out, bones are relational entities, which emerge and take shape through the interactions that take place with them.[4] This transformation of the meanings of bones has, for example, been vividly revealed by the historian of medicine Anita Guerrini regarding the skeletons in the collection of the French anatomist Joseph-Guichard Duverney (1648–1730), which hold a variety of meanings: emotional, moral, religious, medical, and scientific.[5] She has also pointed out that we lack a history of the skeleton in the eighteenth century.[6] Although such a history in beyond the scope of this chapter, the aim here is to understand how a changing idea of aesthesis shaped several types of bone preparations in the eighteenth-century Leiden anatomical collections. The bones and skeletons in these collections of this period had fluid meanings that could change over time, and aesthesis defined much of those meanings and changes. Starting from an iconic skeleton and a wet bone preparation, this chapter shows that what at first sight appears to be one kind of fairly straightforward naturalia—bones and skeletons—were actually artifacts that were defined by aesthesis and which had diverse meanings and uses within one institutional collection, all within a relatively short space of time. But before moving on to these particular preparations, we need to briefly consider what was so far known up to this point concerning these skeleton and bone preparations in the eighteenth-century Leiden anatomical collections.

Approximately 2500 of the more than 8000 preparations in the Leiden anatomical collections today are of bones, and of some of the historical bone preparations we know quite a lot about how they were used and perceived at the time. For example, we know that the collection contained human skulls, collected by Leiden anatomy professor Otto Heurnius (1577–1652) from around

3 Rosner, *The Anatomy Murders*, and Santing, "Andreas Vesalius's De Fabrica," 59: The obvious exception to this would be bones robbed from old graves, yet as these also needed to be cleaned and selected there appears to be a manufacturing and commodification aspect adhering even to them. Excavated bones were not very popular with anatomists, as they were often too badly damaged to be of any use. However, scarcity meant that anatomists could often not be too picky.

4 Hallam, "Articulating bones," 467.

5 Guerrini, "Duverney's Skeletons."

6 Guerrini, "Anatomists and Entrepreneurs," 229.

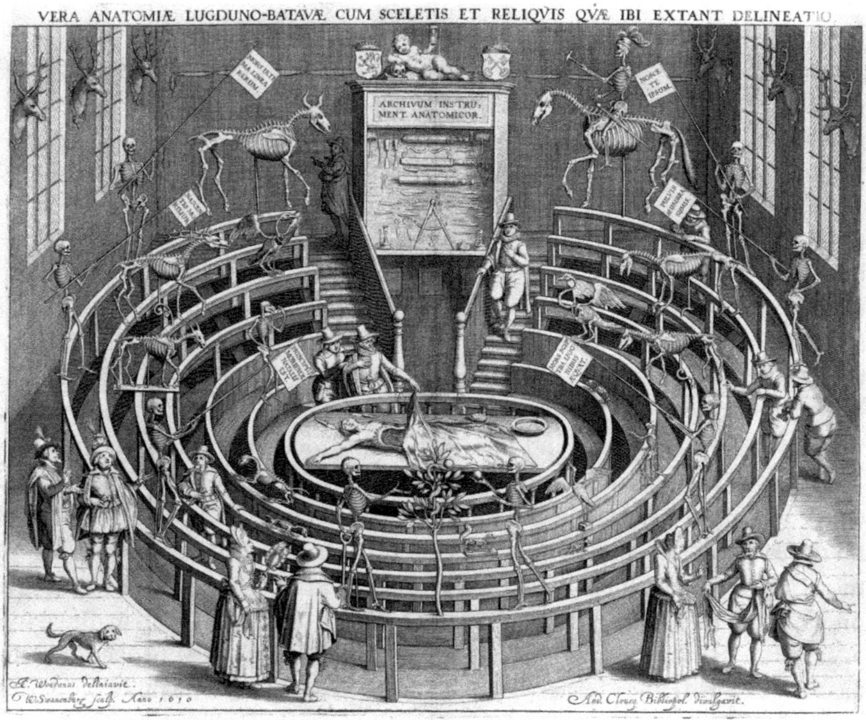

ILLUSTRATION 7.0 *The anatomical theatre of Leiden University, early 17th century. Contemporary engraving by Willem Swanenburgh; drawing by Jan van 't Woud. Around the theater, allegorical displays can be seen, such as the two skeletons in the foreground with an apple tree with a snake in between them. These represent Adam and Eve and the original sin.*

the world, and that by the early eighteenth century the Leiden anatomical theatre was decorated with skeletons of humans and animals.[7] It is also known that S.J. Brugmans' collection of skulls with gunshot and sabre wounds was collected on battlefields in the late eighteenth and early nineteenth centuries to learn more about the impact of external trauma on the skull, and that Gerard Sandifort, Eduard's son, listed and depicted a series of skulls of people from different parts of the world during the last decades of the eighteenth century, as this fitted with the contemporary interest in concepts of race.[8] Yet almost

7 Huisman, *The Finger of God*.

8 Elshout, *Het Leidse kabinet*, 102–3, G. Sandifort, *Museum Anatomicum*, vol. III, IX.

nothing is known concerning the use, making, and collecting of skeletons and bone preparations in Leiden anatomy in the eighteenth century.

Two case studies can start to fill this void. For once, one is not so much a preparation, but illustrations of a preparation by Albinus. This case is an exception in terms of a case study of the materiality of the eighteenth-century Leiden anatomical collections, both because the skeleton that served as its model disappeared, and because it is accompanied by a text in which the mostly tacit aspects of aesthesis are for once voiced quite explicitly. However, the illustrations are similar to Albinus' anatomical preparations in the sense that in both cases he has produced refined, polished anatomical objects that are only distant reminders of the gory, messy reality of anatomical practice. The lost materiality of the portrayed skeleton will serve as a starting point for understanding the hands-on practices, ideals, and bodily involvement of making a bone preparation.

Subsequently, a preparation by Van Doeveren will help to explore thus far virtually ignored differences between the uses and meanings of wet and dry bone preparations throughout the eighteenth century. The elements of aesthesis, which emphasize seeking perfection, gaining knowledge through sensory perception and hands-on work, dealing with disgust, and commodification, influenced all these objects to various extents. This resulted in a wide variety of composition, preservation, and display. However, by the end of the eighteenth century the epistemic culture of aesthesis started to change and disappear, and that too can be understood from the materiality of the eighteenth-century Leiden bone preparations.

Albinus' Perfect Skeleton

For Bernard Siegfried Albinus, professor of anatomy at Leiden University from 1721 to 1770, the human skeleton played a central role in his work. Just how important it was shows from one of the few portraits made of him during his life: in it, a curtain behind him is pulled aside to reveal a skeleton. In 1747 Albinus published his magnum opus: an anatomical atlas of the bones and muscles perspicuously titled *Tabulae Sceleti Et Musculorum Corporis Humani*. Much has been written about this brilliant atlas and the hugely popular plates in it, yet the actual bodies used for the illustrations have until now gone largely undiscussed.[9] Yet it is the actual hands-on laboratory work on the material of

9 Elshout, *Het Leidse kabinet*, Punt, *B.S. Albinus*, Huisman and Luyendijk-Elshout, *De Volmaakte Mens*, Hendriksen, "Une Recherche Commune."

ILLUSTRATION 7.1 *Portrait of Bernard Siegfried Albinus with skeleton in background. Mezzotint by J.J. Haid after a painting of C. de Moor. Reproduced from Punt 1983 (Call number* DOUSA *82 0756). © Leiden University Library 2012.*

the skeleton that was the model for these plates that exemplifies the paradoxical contrast between the corporeal gore and effort of the original materials and the clean and polished final result, in which the visceral materiality of the original has entirely and purposely disappeared. In order to understand Albinus' work on skeletons, we need to know more about contemporary practices of preparing and conserving them. This section will therefore start with an exploration of the practicalities of making dry bone preparations in general, and in other uses of bone, before moving on to the corporeal practices advocated by Albinus in his atlas.

As we have seen, dry preparations of bones and skeletons were very important in the seventeenth-century Leiden anatomical theatre, not least because the skeleton was considered to be the foundational structure of the human body. This also appears from the iconography of the famous 1543 anatomical atlas by Andreas Vesalius (1514–1564), *De Humani Corporis Fabrica*, which can be considered the grand predecessor to Albinus' atlas. In the seventh book, the anatomy of the human body is represented in a series of plates, showing different stages of dissection and thus different layers of tissue and bone. All bodies are positioned in Arcadian landscapes, some with a city or a ruin visible in the distance. The closer to the bones the body is depicted, the emptier, more devoid of human intervention the landscape appears to be, suggesting that the skeleton is the ultimate constitution of man, far removed from the woes of civilization. In the plates of the full skeleton, stripped of all soft tissue, the draughtsman has reverted to the ultimate Renaissance *memento mori* theme: the skeleton is contemplating mortality over a human skull, placed on a tomb.[10] But in order to have these plates made, preparations had to be created.

Before the late seventeenth-century invention of methods for wet preservation of specimens, it was simply impossible to create lasting wet preparations.[11] The first dry bone preparations were either not preparations at all—in the sense that they had not been intentionally preserved but acquired from graves—or preparations made using techniques that were based on natural processes of decay. Wet maceration for example; the cleaning of a skeleton of the last bits of soft tissue by lowering it in a pierced case in running water, or letting it decompose in water that is frequently refreshed.[12] Vesalius in his atlas already recommended the repeated washing of a cadaver with slaked

10 Vesalius, *De Fabrica*. The book, including the plates can be viewed on http://www.nlm.nih.gov/exhibition/historicalanatomies/vesalius_home.html.

11 See chapter 3 for a more extensive discussion of the development of wet preparation techniques.

12 Sturm, "Präparationstechniken Und Ihre Anwendung," 377.

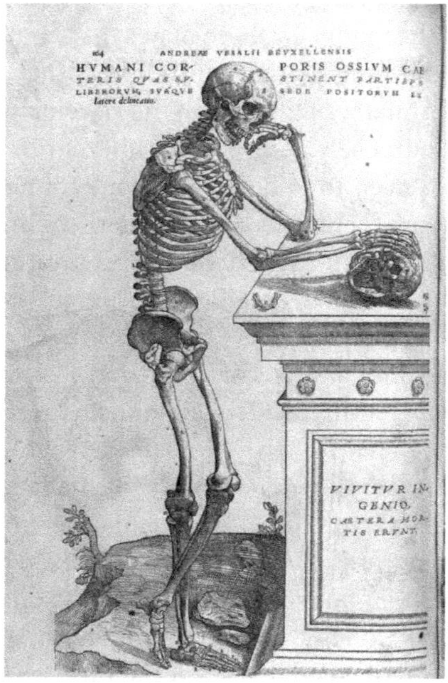

ILLUSTRATION 7.2 *Skeleton contemplating death on skull and tomb, from Vesalius'* De humani corporis fabrica *(Shelf mark Thysia 1622). © Leiden University Library 2012.*

lime and water.[13] Other tested methods of creating bone preparations involved the removal of soft tissue with maggots and by boiling of bodies and body parts.[14] By 1790, Pole still recommended wet maceration as the most effective way of preparing bones and skeletons, although it required patience: three to six months depending on the weather. Of course boiling was quicker, but the downside of that method was that the heat might cause the remaining medullary oil in the bones to penetrate the compact substance and thus give them a "disagreeable oily transparency" which "spoils them of their Beauty."[15] William Hunter too was concerned that just boiling bones would give them a brownish tinge.[16] Moreover, one seventeenth-century handbook writer noted, the marrow containing the medullary oil should not be discarded after extraction, as it

13 Ibid.

14 Ibid.

15 Pole, *The Anatomical Instructor*, 99–100, Lyser, *The Art of Dissecting* (*Culter Anatomicus*), 228.

16 Guerrini, *The Value of a Dead Body*, 257–8.

could be used for other purposes—purposes apparently so commonly known that he thought it unnecessary to specify them.[17]

Yet bones, especially dry preparations of healthy bones and particularly complete skeletons, served primarily as teaching material and mnemonics, as appears from Lyser's 1665 anatomical handbook—a manual reprinted and translated well into the eighteenth century, and recommended to students by Herman Boerhaave.[18] Lyser states that a preserved dry skeleton is the next-best method for studying the "Figure and Shape of the human Bones" after fresh in-situ bones. Throughout the eighteenth century, the general idea of a successful dry bone preparation was that the bones were thoroughly cleaned, attached to one another in the natural order either by the ligaments or in a manner that allowed the user to move the limbs, and that it was as white as possible.[19]

The desired whiteness of bones may seem odd, as bones in the living body are not white at all, but white is traditionally associated with purity, cleanliness and death. The preference for white bones can also be found with Ruysch, who repeatedly mentions whiteness as a desirable characteristic of bone preparations, for example in the preparation of a "very white skeleton of a mole."[20] These rules-of-thumb ensured not only that the skeletons looked beautiful but also that the bones were less prone to decay and thus could be used for the study of normal human anatomy for a long time. How, then, did Albinus use skeletons and bone preparations in his work?

As previously stated, one of Albinus' big projects, and maybe his grandest, was the creation of his anatomical atlas, the *Tabulae Sceleti Et Musculorum*

17 Lyser, *The Art of Dissecting* (*Culter Anatomicus*), 236–7. This is not as odd as it may seem. Richard Sugg, *Mummies, Cannibals And Vampires*: shows that corpse medicine—the use of human bones, as well as human skin, blood, hair, and fat as simplica for drugs- was quite common in Western Europe until the eighteenth century. As the anatomical theatre was one of the few places in the seventeenth century where human remains were relatively easy to obtain, it is not unthinkable that in Leiden too remains that could not be used to create instructive and moralizing decorations for the theatre were sold to apothecaries as simplica.

18 Boerhaave did so in his work on 'studying physics', according to Thomson's preface, to his 1740 translation of Lyser's *The Art of Dissecting* (*Culter Anatomicus*), 5.

19 Lyser, *The Art of Dissecting* (*Culter Anatomicus*), 226, Pole, *The Anatomical Instructor*, 99–100.

20 Ruysch, *Alle Werken*, 722. Whitening the bones may also have been a successful strategy in the commodification of the skeleton. It is not hard to imagine that a clean-looking, white skeleton is both less disgusting to handle and easier to trade than a grubby yellowed one. Hallam, "Articulating bones," 482, gives more examples of a preference for white bones among anatomists.

Corporis Humani, about which he started thinking as a student. He wanted to create an anatomical atlas that was more accurate than any of the atlases available at the time.[21] This was not an unusual ambition for an anatomist—many created their own atlases, prestigious, expensive volumes, with the help of skilled craftsmen such as draughtsmen and engravers. In fact, there was a strong reciprocity between iconography and anatomy in early modern Europe. As Hallam has pointed out, mori objects like skeletons carved in ivory were often created by the same craftsmen who sculpted miniature anatomical manikins for use as 'instructive tools' by doctors in medical training in northern Europe.[22] These craftsmen in their turn obtained their knowledge of human anatomy in the anatomical theatre, from the dissections and the human skeletons on display there. In mori objects, such as ivory skeletons, vanitas-paintings, and the skeletons on display in the early Leiden anatomical theatre, anatomical realism and *memento mori* symbols were fused. This led to layered objects, which were both reliable instruments for anatomical instruction and subjects for spiritual contemplation.

However, Albinus was not very interested in memento mori symbols—he wanted the atlas to be as clean-cut and clear as possible. The Arcadian landscapes in the background of the plates, and the rhinoceros Clara were only added at the instigation of his draughtsman, Jan Wandelaar, who thought they would benefit the sense of scale of the skeletons and ecorchés.[23] As Albinus was so obsessed with perfection, the production of the atlas occupied him for the rest of his life. Only the parts dealing with the bones and the muscles were finished when he died; a few sketches for the part on the internal organs remain. The creation of the atlas, especially the technicalities of the production of the plates, and the details of the cooperation between Albinus and his draughtsman, has been discussed extensively elsewhere.[24] But in order to make the plates, Albinus obviously needed a body—more than one in fact, as bodies deteriorate quickly—, and he did not only depict the full skeleton, but also all the separate bones and the different layers of muscles. What Albinus recounts of the process of selecting and maintaining bones and skeletons for the pro-

21 Punt, *B.S. Albinus*, 15.

22 Hallam and Hockey, "Death, Memory and Material Culture," 65.

23 Huisman and Luyendijk-Elshout, *De Volmaakte Mens*, 14. For more on Clara, see Verheij, *Op Reis Met Clara*.

24 See Elshout, *Het Leidse kabinet*, 53–5, Punt, *B.S. Albinus*, 14–68, Huisman and Luyendijk-Elshout, *De Volmaakte Mens*, Kemp, " 'The Mark of Truth'," 107–12, Cunningham, *The Anatomist Anatomis'd*, 253–62, Hendriksen, Hendriksen, "Une Recherche Commune," 22–34.

ILLUSTRATION 7.3 *Skeleton, from B.S. Albinus'* Tabulae Sceleti et Musculorum Corporis Humani. (*Shelf mark* PLANO *54 D 1: 1*). *© Leiden University Library 2012.*

duction of his atlas is very enlightening. Paradoxically, the lost materiality of Albinus' model skeleton sheds light on the corporeal and sensory involvement of anatomists with the materiality of the Leiden collections in a broader sense.

In the preface to his anatomical atlas, Albinus describes the difficulties he encountered with his material during the production of the plates. The first part of the atlas was on the skeleton, as Albinus considered the skeleton to be the foundational structure of the human body. He wanted to reach a level of detail unprecedented in other anatomical atlases: first the full skeleton had to be depicted from three different sides (front, back, and sideways), and then all the bones had to be depicted and described individually.[25] For the depiction of the full skeleton, Albinus wanted a specimen that was fully grown, a state that he defined by the epiphyses being fully developed and attached to the bones. This indicates that he considered the young adult human skeleton to be perfect; childhood was a formative but imperfect phase, and after a certain age, decay set in. This attitude towards childhood was common in anatomy until the early eighteenth century. However, unlike his seventeenth-century predecessors, Albinus was also interested in the formative stage, as we will see in the next section.

25 Albinus, Albinus's Anatomical Figures, preface p. v.

For the depiction of the anatomy of the human body, however, Albinus wanted an adult skeleton. He eventually found an ideal one, one that met all his criteria of elegance and beauty:

> It was of the male sex, of a middle stature, and very well proportioned; of the most perfect kind, without any blemish or deformity, either as to the bones themselves, or their connexions with one another. And as skeletons differ from one another, not only as to age, sex, stature, and perfection of the bones, but likewise in the marks of strength, beauty, and the make of the whole; I made choice of one that might discover signs both of strength and agility; the whole of it elegant, and at the same time not too delicate; so as neither to shew a juvenile or feminine roundness and slenderness, nor on the contrary an unpolished roughness and clumsiness; in short, all the parts of it beautiful and pleasing to the eye.[26]

It sounds too perfect to be true, and it actually was, Albinus admitted. But that was the good thing of making an anatomical atlas: unlike in a preparation, where the anatomist could only correct blemishes and irregularities to a certain extent, in the plates he could correct every single one of them. To Albinus, this was not some kind of artificial beautification. In fact, he wrote: "by this method it is possible to express the perfection of nature, at least to come very near it."[27]

So Albinus had his (almost) perfect skeleton, but having the drawings made was a lengthy, arduous process. Albinus wanted to put up the skeleton with the ligaments and cartilage attached, in order for it to appear as natural as possible, and he draped it in a rather complex position, using chords to keep it that way. But because he kept the ligaments and cartilage, he could not clean the bones with the traditional maceration methods, nor did he want the whole to dry out, as he felt this would negatively affect the natural looks of it. This meant the skeleton was prone to decay, especially as the room in which the draughtsman Jan Wandelaar was working had to remain slightly heated. This

---

26 Ibid., preface p. xiv. Tour guides in the Leiden Museum Boerhaave have suggested that the arms of the skeleton in Albinus' plates are too long and that 'a specialist' has declared that they are not human but chimpanzee bones. However, it is quite unlikely that Albinus would deliberately have used chimpanzee bones to depict his 'homo perfectus'; it seems to make more sense that the relatively long arms are either the result of some kind of optical distortion in the drawing process or of Albinus' peculiar ideas about the perfect proportions of the body.

27 Albinus, *Albinus's Anatomical Figures*, vi–vii, xiv.

had nothing to do with considerations about cold hands: Albinus had come up with the luminous idea of positioning a naked, very skinny man next to the skeleton in the same position so as to help Wandelaar get it right, but the man demanded for the fire to be kept on at all times—not surprising as the work took place during winter months. In order to make sure the skeleton lasted long enough to finish the drawings, about three months, Albinus had to tend to it continuously:

> ... it was necessary to take care, that the skeleton should not suffer from drying or putrefaction, nor were hindered from going on from the same causes. Sometimes then, when it dried too much, I moistened it with water, and cutting the ligaments, poured it into the joints, in order to preserve the cartilaginous crusts; and again, when the putrefaction was to be checked, I sprinkled it with vinegar; wrapped it up in the night time with paper, and cloths dipped in the same liquor; and poured some of the same wherever there was room for it.[28]

Here we once again see the curious mix of wanting to find and display perfection and elegance in anatomy and the undoubtedly stinky, sticky, disgusting messiness of anatomical practice. It could be argued that the world was a dirtier, smellier place in the eighteenth century, and that therefore a chore as described above would have filled Albinus with less disgust than it would us. Alberti has argued that the wonder and disgust experienced by people in other times differs from our own experiences upon encountering the same situations and objects.[29] However, Albinus was allegedly famous not only for his eloquence and preparation skills, but also for his capability of never soiling his sleeves while dissecting, suggesting he was rather fond of order and cleanliness—a characteristic that seems almost incompatible with repeatedly moistening a decaying skeleton.[30] Other eighteenth-century accounts of experiencing disgust upon witnessing the inside of the human body make it clear that this is a form of core disgust that had to be overcome by anatomists back then just as much as today. The lawyer and poet Willem Bilderdijk for example, stated in a 1783 thesis on the beauty of the human body:

---

28 Ibid., x.

29 Alberti, *Morbid Curiosities*, 195.

30 Von Haller, *Bibliotheca Anatomica*, vol. II, 126, lists Albinus as the first of what he calls the 'perfect anatomists', a title he unfortunately does not elaborate upon.

> Because it is in the first place the idea of the soulless, the dead, of the corpse itself, which (notwithstanding its appearance) both in the dissected body and in all other corpses evokes disgust. (…) Added to this, second, is the realization of abuse, which we receive from the contemplation of cut, slashed, skinned and maimed corpses. The idea of the human body is too closely connected to our idea of ourselves not to cause a comparison and a transition of ideas, and can evoke a kind of pity, which sometimes can move the intestines, and which is no less powerful because it is unreasonable.[31]

Once we are aware of these tendencies, Bilderdijk continues to argue, we can set them aside and get used to the seeing of, and dealing with, death bodies, and continue to seek the beauty within their structures.[32] Yet even for an experienced anatomist like Albinus, who was aware of the general human tendency to experience the anatomized body as a disgusting reminder of our own vulnerability and mortality, handling corpses cannot have been a walk in the park. His awareness of the disgusting aspects of anatomy is expressed in his preparations and his atlas images, in which he tried to—sometimes literally—cover up the reminders of human vulnerability and mortality and to bring out the ideal, beautiful human body. Moreover, Albinus, the man of whom we might think is the ultimate example of anatomical sprezzatura, judged by his atlas plates and preparations, happily admitted to the bodily discomfort of creating these images and objects in his writing, speaking of the 'pains, labour and expense' they required.[33]

In Albinus' work on the creation of his anatomical atlas we once again see all the aspects of aesthesis. He sought an ideal, beautiful body, using his sensory experiences of it to create images of an even more ideal body, something that could only be done through a lot of practical, hands-on work. That work was often dirty and arduous, yet dealing with the disgust it provoked, Albinus admitted, was a necessary exercise in creating anatomical objects that opti-

---

31 Bilderdijk, "Antwoord", appendix v, 131–2: "Want, in de eerste plaats is 't het denkbeeld van 't ontzielde, van 't doode, van 't lijk-zelf, dat (ongeacht deszelfs gedaante) in 't ontlede Lichaam even als in alle lijken den afkeer opwekt. (…) Hier bij koomt, ten tweede, 't besef van mishandeling, het welke wij door de beschouwing van opengesneden, gekorven, ontvelde, en verminkte lijken ontfangen. Het denkbeeld van 't menschelijk lichaam is te na aan het denkbeeld van ons-zelven verbonden, om hier geen vergelijking en overgang van denkbeelden te veroorzaken, en ons een sort van meêlijden te doen ontstaan, dat somtijds de ingewanden beroeren kan, en niet minder kracht heeft om dat het onredelijk is."

32 Ibid., 132–3.

33 Albinus, *Albinus's Anatomical Figures*, xxiv.

mally showed the perfection of the human body. It is fascinating to note that although he readily admitted to the material and visceral reality of the skeleton that was used as a model for his atlas, Albinus did not care to mention what happened to it eventually. Unlike Vesalius, who created skeleton preparations as well as an anatomical atlas to teach anatomy, Albinus purposely made the atlas the sole product of his project.[34] He got rid of the material, messy reminder that was the almost-but-not-entirely perfect skeleton, maintaining only the clean atlas plates.

The absence of the materiality of the model skeleton first seems to distinguish it ontologically from the other cases in this book. Yet Albinus' attempt to map his physical involvement with the material he used in the plates and preface of his atlas shows that he felt it was important to inform the user that although the original skeleton was lost, this was still the elegant product of his own sensory perceptions, dealings with disgust, and hands-on skill. And it was not in vain: like Albinus' preparations, the results of his visceral efforts were stunning and highly successful. The expensive atlas and its plates were a popular and much-plagiarized collector's item, and Albinus' preparations were equally admired and copied, as shows from a similar wet bone preparation by Van Doeveren, using the style and technique of his teacher.[35]

Wet Bones, Dry Bones: Osteogenesis and Ossa Morbosa

It has long been known that the eighteenth-century Leiden anatomical collections contained both wet and dry bone preparations; both kinds are listed in eighteenth-century catalogues of the anatomical theatre and in the 1793 first volume of the *Museum Anatomicum*. It is not hard to imagine that dry and wet preparations both have their pros and cons in terms of preservation and handling, but why exactly was a particular method chosen for a particular preparation? What role did aesthesis play in these choices? In this section, I use a wet and a dried preparation to trace the development of techniques of bone preparations in Leiden in the course of the eighteenth century. Based on that history, I will argue that specific methods of preservation suited different yet related fields of anatomical inquiry, leading to the simultaneous use of various preservation methods in the Leiden anatomical collections, but noting that in

34 One of the skeletons prepared by Vesalius is still on display in the Anatomisches Museum Basel (Switzerland).

35 Ad0003, a vertebral column of a six months old foetus, from the collection of Van Doeveren.

keeping with aesthesis, considerations about the elegance and perfection of the resulting preparations always played a part.

In the collection of Wouter van Doeveren we find a number of wet bone preparations like this one, a vertebral column of a six months old foetus.[36] It is a nice preparation, showing very well the still largely cartilaginous structure of the vertebrae. Six other preparations of bones of foetuses, a newborn, and an infant remain in the Van Doeveren collection, forming the majority of his few bone preparations. The need for dry bone preparations seems obvious, but why would one want to make a wet preparation of bones? This can be understood from a shift in anatomy from a search for the normal, average, adult skeleton to questions of generation. For example, Lyser still recommended an adult skeleton for the study of anatomy, as he believed these were more durable and more perfect than the skeletons of children and even adolescents:

> We indeed often see Infant Skeletons in Anatomical Theatres and the Repositories of Anatomists; but these serve more for Ornament than real Use, and are made up in a quite different Manner to adult Skeletons.[37]

This suggests that although he was aware of the fact that the proportions in a child's skeleton are very different from those of an adult skeleton, he was not particularly interested in those differences—he appears to have thought of childhood as an imperfect pre-stage of adulthood—and even thought the preservation of children's skeletons a bit whimsical. That attitude would change in the eighteenth century with the growing interest in the source and force of life, although the entire adult skeleton, as well as separate bones, healthy and diseased, also remained objects of study throughout the eighteenth century.

By the late seventeenth century, the Briton Clopton Havers (1657–1702) had discovered that bones are full of microscopic veins and canals. This led Boerhaave to develop the idea that bones were formed through the apposition of membranes.[38] Probably inspired by these developments, Rau first developed the technique of creating nearly transparent wet bone preparations, particularly of foetal and infant bones, in the late seventeenth or early eighteenth century.[39] Why did Van Doeveren create these few, particular bone preparations

36 Ad0003.

37 Lyser, *The Art of Dissecting* (*Culter Anatomicus*), 227.

38 Boerhaave et al., *Boerhaave's Orations*, 315.

39 Elshout in the 1950s catalogue of the anatomical collections noted on Van Doeveren's wet bone preparations that the manner of preparing is different from that of Rau and Albinus, that it somehow looks more modern.

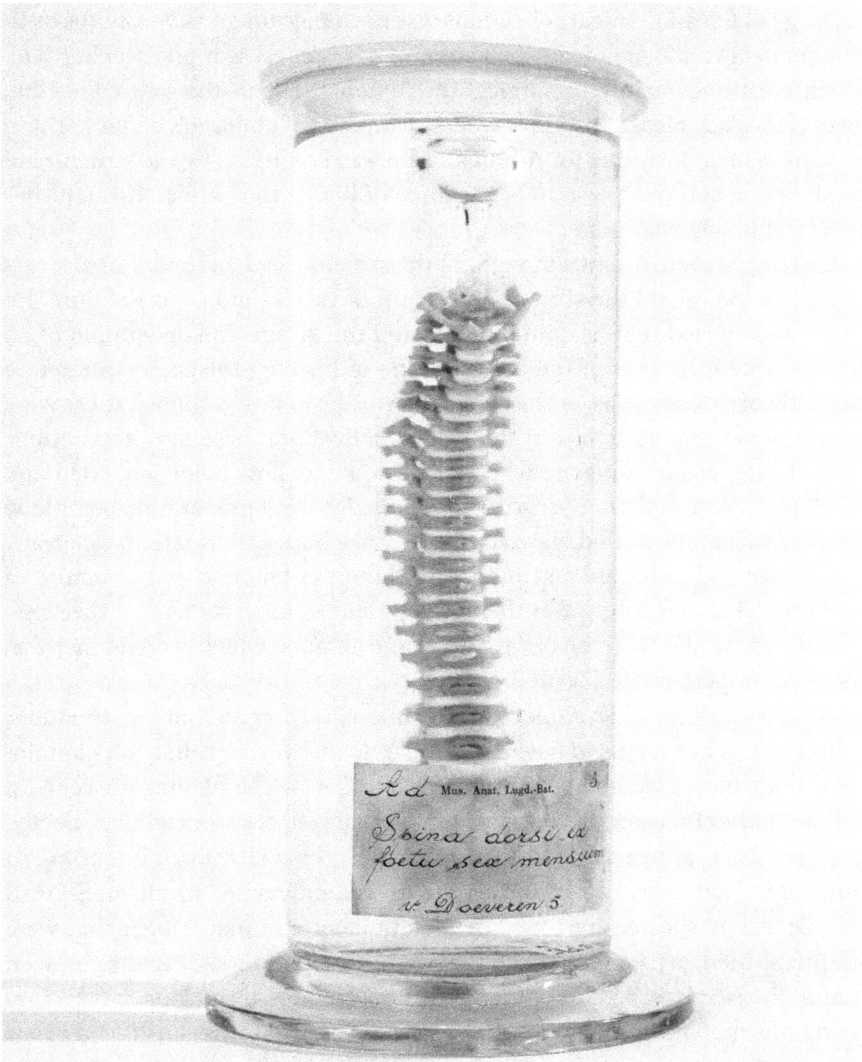

ILLUSTRATION 7.4 *LUMC Ad0003 Wouter van Doeveren (1730–1783) A vertebral column of a six months old foetus, preserved wet. © Anatomisch Museum LUMC 2012. Photographer: Arno Massee.*

so many years later? In order to understand that, we have to look again at the man who first taught him anatomy when he was a student in Leiden: Albinus. When Albinus was appointed as lector of anatomy in 1719, the status of the existing Leiden anatomical collections was to change for good.[40] Albinus dutifully made a catalogue of the collection of his predecessor and teacher Rau, which consisted mainly of surgical instruments. But in the Rau collection, there were also at least thirty-five wet preparations of children's bones.[41] These must have been inspiring for Albinus, as he was fascinated by the structure of bodily tissue and the question of the force of life, of that which distinguishes living from lifeless matter.

As we have seen, Albinus thought of the skeleton as the foundational structure of the body, and the structural element of the skeleton was the fibre. He therefore believed that he could understand the nature and generation of all bodily tissue from that of the bones.[42] One of his big projects was therefore the study of osteogenesis, or the generation of the bones. Although there were several bones and skeletons in the Leiden collection, these dry preparations were of little use to someone who wanted to know how bone generated and developed. Albinus therefore further developed the wet preparation technique for bone preparations that allowed him to make bones transparent, especially of foetuses and children, and to clearly show the tendons and structure of the bone. This new interest in the development of bone structures gave new importance to the skeletons of foetuses and infants, which were earlier dismissed as imperfect, irrelevant or whimsical.

Albinus must have developed this technique relying on Rau's instructions, as his old teacher seems to have been the only one with a collection containing similar preparations in the first two decades of the eighteenth century. Albinus' father for example, Bernard Albinus (1653–1721), still mainly collected dried bones, as appears from the account given by a German student in 1711 (when B.S. Albinus was only fourteen). This student, named Erndl, mentioned that 'Dr. Albin' showed him "two Articulations of a human Finger that were petrified [dried, MH], in which all the Apophyses [epiphysis] were to be seen accurately express'd."[43] So apparently it was possible to make a satisfactory dried bone preparation showing the epiphysis, but what was considered a flaw in dry bone preparations, an oily transparency, turned out to be a valuable asset in a wet bone preparation. This transparency allowed for the study of the

40 Also see chapter 3.

41 Albinus, *Index Supellectilis Anatomicae*, 18–21.

42 Albinus, *Icones Ossium Foutus Humani*, 6., Punt, *B.S. Albinus*, 88.

43 Erndl, *The Relation of a Journey*, 59. An epiphysis is a part of a bone that ossifies separately and later becomes attached to the main part of the bone, or the end of a long bone.

internal structure of the bone without having to have access to a fresh body for dissection, and actually provided clearer views of internal bone structures than fresh bones.

Apart from these practical benefits, Albinus' wet bone preparations served as the material, anatomical proof in a physiological debate. Around 1740, the theory of preformation was still generally accepted. This theory held that the human being does not gradually emerge from undefined material, but that is already rudimentarily present in either the male semen or the female ovum—the advocates of these locations of origin were called animalculists and ovulists respectively.[44] From observations of embryological development, Albinus learned that the extremities developed from the embryo as from an immobile bud. Combined with the preparations he created of foetal and infant bones, this led him to argue that the embryo was not a preformed human, but only an imperfect rudiment of a human, developing slowly under the influence of the enormoun (ενορμουν), a sort of driving force of the bodily fluids derived from nutrition.[45] The bones emerged from the gelatinous embryo and gradually became more solid, until, after about thirty years, they had reached their perfect state. After that, their perfection decreased again.[46] So where Lyser had simply denounced preparations of children's skeletons as 'imperfect,' and thus not relevant almost a century earlier, Albinus prepared the imperfect skeletons of foetuses and infants precisely to understand the genesis and development into perfection of the human bones.

These unique wet bone preparations, together with Albinus' atlas project, quickly brought him lasting international recognition as a specialist in the subject of osteology and osteogenesis. The German physiologist Albrecht von Haller (1708–1777), who met Albinus when he traveled through Holland in 1725–1727, stated that Albinus' bone preparations of foetuses were better than those made by Ruysch, and much more detailed:

> He [Albinus] prepared the bones of a child the size of a big nut in such a way that one can discern all the bones of the hands, and the head with the dilated mouth . . .[47]

---

44 Punt, *B.S. Albinus*, 102.

45 On *enormoun* in eighteenth-century physiology also see Van der Korst, *Een Dokter van Formaat*, 34–6.

46 Punt, *B.S. Albinus*, 102, Box MS 1741–1742, f.576–7.

47 Von Haller, *Albrecht Hallers Tagebücher*, 93: "In den Knochen aber eins Kindgen, wie eine große Nuß groß, so zubereit, daß man alle Knochen der Händen, dem Kopf und aufgesperreten Mund (. . .) an ihrem Orte kan liegen sehn."

Albinus' English friend and colleague, Robert Nesbitt (?—1761), in the dedication to his 1736 book on osteogenesis, once again stressed how important it is that the transfer of anatomical knowledge is supported by either demonstration or preparations, and mentions that he took Albinus' preparations as an example for his own:

> ... I was obliged, in order to prove to my hearers the truth of my descriptions, to make a large number of preparations very different from what I had ever seen in any anatomical museum, except for that very curious one of professor Albinus at Leyden. When you examined those preparations, you was pleased to approve of them; because each part, I designed to describe, was preserved in natural shape and colour.[48]

The reason these preparations kept their natural shape and colour was due to the new wet preservation technique Albinus and now also Nesbitt used; Nesbitt stated that skeletons drawn from dried preparations look very different from their 'natural state.'[49] Here he indeed had a point. Although a thoroughly cleaned, dried, and whitened skeleton was a useful long-lasting teaching aid, bones are not naturally white as they are permeated with veins and nerve ends, and they shrink and discolour from the treatment that is necessary to preserve them dry. As we have seen in the previous section, Albinus went to great lengths to prevent that from happening to the skeleton he used for his atlas. With Albinus' perfecting of the wet bone preservation technique, he once again proved that he was a master at combining sensory experiment with the creation of elegant, refined preparations in which structures remained as close to their original state as possible. Of course, the preparation of the spine was actually not made by Albinus, but by his student and successor Van Doeveren. This shows that Albinus remained an example for future generations of anatomists. For Van Doeveren, wet bone preparations may also have played a role in his understanding of the development of monstra, although he does not refer to it explicitly in his writings.[50]

However instructive wet bone preparations were for the study of the development of bones, they were not suitable for all aspects of osteology. Not even Albinus, the man who was famous for his refined preparations of perfect anatomy, could steer clear of pathology completely, as appears from this prepara-

---

48 Nesbitt, *Human Osteogeny Explained*, v–vi.

49 Ibid., vii.

50 For more on Van Doeveren's ideas about the formation of monstra or birth defects, see chapter 5.

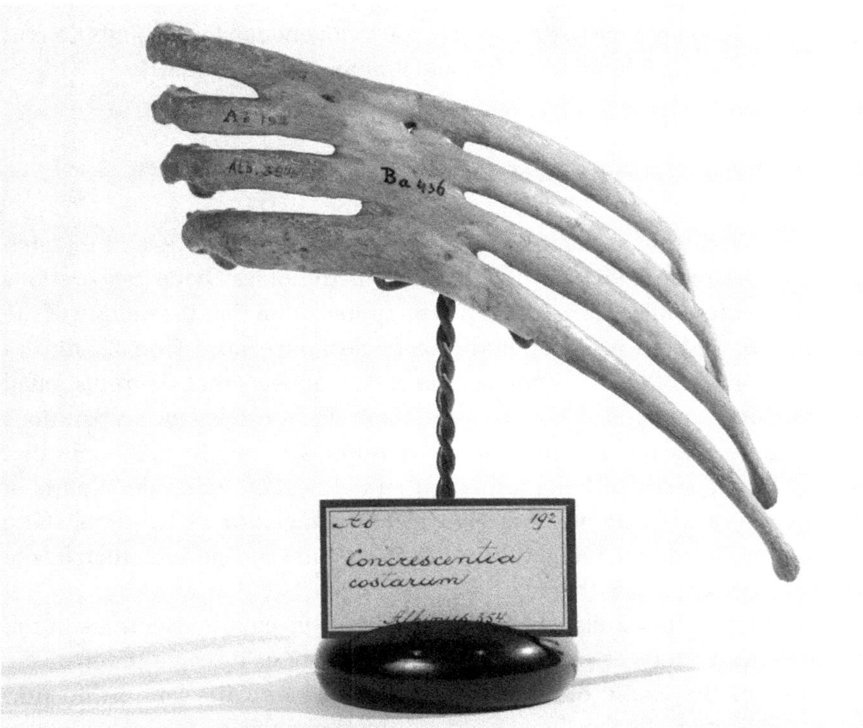

ILLUSTRATION 7.5 *LUMC Ab0192. B.S. Albinus (1692–1770) Costa synostose. Four right ribs of a human, fasciated in the middle. Between the first ribs two small holes are visible. © Anatomisch Museum LUMC 2012. Photographer: Arno Massee.*

tion of fasciated ribs.[51] As we have seen, Albinus spent almost all his time and energy on his atlas, and on research concerning *vis vitalis* and the anatomy and physiology of the senses. If he collected pathological specimens, he appears to only have done so if they might tell him something about his primary interests: for example preparations of concrescences and degenerated body parts may help understand the power of *vis vitalis*. Yet such preparations remained peripheral elements in Albinus' work. In his vast collection, only seven pathological preparations, of which three are of diseased bones, were listed in 1793.

51 Ab0192 costa synostose. Four right ribs of a human, fasciated in the middle. Between the first ribs two small holes are visible. Sandifort, *Museum Anatomicum*, vol. I, 180., vol. II, 84, and Sandifort, *Supellex anatomica*, 228.

As we have seen previously, that was certainly not enough for the ambitions of the new professors Eduard Sandifort and Wouter van Doeveren.[52]

Changing Uses, Fluid Meanings

Sandifort's and Van Doeveren's ambitions in anatomy teaching and research heralded the re-appreciation of a number of the oldest bone preparations from the Leiden collections. It is in this re-appreciation that the fluidity of the meanings of the bone preparations becomes clear once more. From Sandifort's 1793 first volume of the *Museum Anatomicum*, it appears that Heurnius' small selection of (fragments of) exotic skulls, Rau's skulls, osteogenic preparations and a couple of healed fractures, as well as Albinus' three diseased bones that he used in the study of bone formation and *vis vitalis*, were valued parts of the anatomical collections by the late eighteenth century.[53] However, these preparations were used differently by Sandifort and Van Doeveren than by the predecessors who made them.

For example, the skulls and skull fragments collected by Heurnius in the seventeenth century, as either a curious pathology or a scarce proof of the existence of the exotic other, were incorporated into the late eighteenth-century collections by Sandifort as just another example of a particular disease or inhabitant of a certain geographical region. Moreover, the few deformed bones that served Albinus well as comparative material in his study of osteogenesis were not sufficient as teaching material for the new classes on pathology. The Leiden professors therefore also contributed to the collection of bone preparations themselves. As we have seen, Van Doeveren added a number of osteogenic preparations to the collections, and Eduard Sandifort himself compiled a collection of over one hundred human bone preparations, mostly pathological.[54]

---

52 Also see chapter 5.

53 These anatomists are all mentioned in the introduction, and their human bone preparations are listed in the *Museum Anatomicum*. This is a clear indication that they were used and appreciated by the then anatomy professors, as Sandifort also made a separate list of objects and preparations present in the anatomical theatre that he deemed no longer suitable for an anatomical collection, such as an Egyptian mummy collected by Heurnius, and some bottles with insects from Albinus' collection. This list from 1798 is kept in the Archives of the Curators, no. 228. A transcript can be found in Witkam, "Over De Anatomieplaats," appendix 38, 62–5.

54 LUMC catalogue. G. Sandifort, *Museum Anatomicum*, "Preface," and Huistra, "Preparations on the Move:" After the Royal Decree of 1815, which determined that all Dutch universities

This reinterpretation of existing collections and the speedy addition of large numbers of pathological preparations coincided with a shift in the fields of anatomy and physiology described by Andrew Cunningham. Cunningham argues that before circa 1800, anatomy was a 'discipline of the sword', that is, a practical, hands-on, experimental discipline, whereas physiology was the 'discipline of the pen'; a philosophical, largely theoretical discipline. In the last decades of the eighteenth century, this division started to shift. Slowly but steadily, anatomy, although it remained relatively hands-on, became less experimental, while physiology became increasingly experiment-based, and eventually the two merged in the new discipline of experimental physiology.[55] This implies that as the epistemic culture of anatomy was changing, the concept of aesthesis is no longer a valid heuristic instrument in describing and understanding it. These changes also became visible in the Leiden curriculum and collections. For example, the teaching methods and topics of the Leiden anatomy professors changed. Although students still had to perform dissections and make preparations themselves to learn anatomy, handbooks became increasingly important as learning aids.[56] Moreover, as we have also seen above, whereas the understanding of disease had previously been a largely physiological, theoretical affair, in the course of the eighteenth century pathology increasingly became a part of anatomy teaching. By the time Sandifort and Van Doeveren were professors in Leiden, they taught their students pathology both through dissection and preparations, and no longer just as a part of theoretical lectures on physiology.[57]

These developments meant that certain aspects of aesthesis inevitably assumed a different role in anatomy: hands-on sensory experience remained essential, yet searching for elegance and perfection in anatomy and proving one's elegant skills as an anatomist became less important. The latter also had to do with the fact that the art of making anatomical preparations had become much more common than it had been at the start of the century. Preparation handbooks had become more detailed and more commonly available, and there were simply more people around who had had sufficient on-the-job

should have substantial pathological teaching collections, the Leiden bone collections would receive extensive additions with the acquisition of the collections of Bonn and Brugmans in 1822 and 1819 respectively.

55 Cunningham, "The Pen and the Sword."

56 For the increasing importance of anatomical handbooks in teaching, also see Richardson, *Death, Dissection, and the Destitute.*

57 Molhuysen, *Bronnen*, vol. VI, 215.

training.[58] This also appears from the speed with which bone preparations were added to the Leiden collections. The increased need for pathological bone preparations combined with the more widespread knowledge of preparation techniques meant commodification of the body became both easier and more important in late eighteenth-century anatomy.

This did not mean, of course, that making preparations had become less messy or disgusting, and although many Leiden anatomists experimented in their own homes too, the facilities housed in the same building as the theatre, the Faliede Bagijnchurch, formed a welcome escape.[59] This building thus remained a site where many preparations were made and displayed throughout the eighteenth century. Anatomy professors often kept at least a part of their preparations in their homes and private offices, but as soon as a collection was donated to, or purchased by, the university, the preparations were housed on the premises of the anatomical theatre—either in the theatre itself or in adjacent rooms.[60] In the course of the century, several extensions and refurbishments were carried out, according to the wishes and needs of the anatomy professor.[61] A stable factor was that the facilities for making preparations were mainly in the cellar of the anatomy building, out of sight of the common visitor, although the downside of this was that unpleasant fumes could travel up to the theatre. Professor Eduard Sandifort was so fed up with the stink in 1797 that he had a new extension added to the cellar, where he moved the water pump and sink, and installed a new maceration coffin, clad with lead. Three years later, at the turn of the century, a new hearth was installed for the benefit of making preparations and performing dissections.[62]

These investments in facilities for making anatomical preparations show that by the end of the eighteenth century dealing with the disgusting aspects of doing anatomy still raised concerns for the anatomy professors. Moreover, it shows that it was still important for the Leiden anatomists to gain knowl-

58 For example, compared to Blankaart's late seventeenth-century *Anatomia Reformata*, Pole's 1790 *The anatomical instructor* is already much more detailed and gives clearer instructions on how to make a preparation. Hyrtl's 1865 handbook *Handboek Der Praktische Ontleedkunst* gives even more precise recipes for injection mass et cetera.

59 Molhuysen, *Bronnen*, vol. V, 19. From the acts of the university curators we learn that B.S. Albinus in 1725 asked and gained permission to lecture on anatomy in the Auditorium of the Falide Bagijnkerk (the room underneath the anatomical theatre), because the crowd was too big for his house. It is striking that he shared the room with the English church council.

60 Witkam, "Over De Anatomieplaats," 24–39.

61 Ibid.

62 Ibid., 27–9.

edge based on their own experiments and sensory perceptions. What Sandifort had stressed in his inaugural lecture in 1772, namely that practical medicine could only be learned through dissection of both the healthy and the pathological body, apparently still rang true for him almost three decades later.[63] Yet the quest for elegance and beauty became less distinct in the anatomists' work, and although practical measures were taken to deal with the disgusting aspects of anatomical practice, deformity and pathology were increasingly considered accepted components of the changing disciplines of anatomy and physiology. Some of the aspects of aesthesis were thus waning, whereas for others the importance was growing, or at least changing, within the epistemic culture of anatomy by the end of the eighteenth century. This also means that the concept of aesthesis becomes less suitable for understanding the anatomical preparations resulting from that culture.

This chapter has shown that the appearance of different kinds of bone preparations in the Leiden anatomical collections throughout the eighteenth century—skeletons, wet bone preparations and eventually a re-appreciation and extension of the collection of dry pathological bone preparations—fitted in with the contemporary research interests of the Leiden anatomists, and resulted from the epistemic culture of aesthesis. Although the kind of preparation that was most desired changed over time, until late in the eighteenth century aesthesis was important in the creation of all kinds of preparations, and older preparations were often fairly easily invested with new meanings and uses within the collections. 'Whimsical' early eighteenth-century infant skeleton preparations gained new meanings in the light of Albinus' theory of osteogenesis, and the wet and dry preparations Albinus used to understand the formation of bone, as well as some of the oldest dry bone preparations that were collected initially as curiosities, in turn were incorporated by Sandifort in his pathological bone collections to help understand bone diseases and malformations. These bone preparations were fluid objects; sometimes used for teaching, at other times for research material, and sometimes both concurrently.

For most of the eighteenth century, seeing with one's own eyes, feeling with one's own hands, searching for and appreciating the perfection and elegance of nature, commodifying and objectifying the human body into a preservable, portable object of study, skill and admiration, all remained determining

63 Sandifort, *Oratio De Circumspecto Cadaverum*, 6: "Haec practicae medicinae Gubernatrix. Hac ad intimosabditosque organorum recessus paratur via, funestae mroborum exponuntur clades, tutior patet curandi methodus, & accuratior instituitur prognosis, ubi fatorum mandata rescindere non licet."

factors for the creation and collection of preparations for the Leiden anatomists. By the late eighteenth century, the Leiden anatomists, particularly Eduard Sandifort, realized very well that they were indebted both materially and theoretically to their predecessors. Yet something did change in the aesthesis of the Leiden anatomists. Although the reliance on sensory perception and a certain hankering after the discerning of beauty and elegance in human anatomy were still important, as shows from Van Doeveren's wet bone preparations, the new volition to compare various cases or preparations in order to gain a systematic understanding of pathology started to become more influential, although the preference for personally dissected and preserved preparations remained, as shown from the investments in the preparation rooms. An anatomist still had to learn from doing with his own hands, seeing with his own eyes, overcoming the stink and gore of the dissection room to create elegant preparations that served his research and teaching activities. However, institutional collections became increasingly important in research and teaching,[64] and it was now perfectly acceptable for an anatomist to also use preparations made by predecessors in his research and teaching—a tendency that would only become stronger in the nineteenth century.[65] This shows that the epistemic culture of aesthesis in anatomy was waning by the turn of the century.

The analysis of specific examples of eighteenth-century Leiden bone and skeletal preparations, set against the background of such preparations in Leiden before 1800, has demonstrated a number of things. First, that skeletons and bones in anatomical collections are never just that: they are preparations, painstakingly dissected, cleaned, and preserved either wet or dry. More often than not skeletons were severed into individual bones, or a certain bone formation was separated, like the trunk, pelvis, or a joint. The parts, together making up what Alberti calls dividuals, are much more important for this eighteenth-century anatomical collection than entire skeletons. This seems to indicate that at least within the realm of anatomical collections, focusing on the history of human bones is more fruitful than focusing on the history of the human skeleton. Second, we have seen that whereas the complete skeleton in the seventeenth-century anatomical theatre served multiple purposes—namely those of empirical educational instruments as well as memento mori—the eighteenth-century Leiden bone and skeletal preparations, although also

64 Huistra, "Collecties Op College," 29.

65 The extension of institutional collections was also influenced by political issues, see Huistra, "Preparations on the Move." Alberti, "Owning and Collecting Natural Objects:" In England, a similar shift from privately to institutionally owned collections was visible in the nineteenth century.

human, were of a very different nature. The complete, average human skeleton remained a basic commodity in anatomical collections throughout the eighteenth century, yet in the Leiden collections during this time specific parts of the skeleton, both healthy and diseased and prepared in distinct ways, served the various focus areas within the collections such as osteogenic research and the study of pathology.

Moreover, the examples in this chapter illustrate that some preparations lost part of their meanings and were invested with new meanings and uses within the collections—such as infant skeletons, old skull fragments, and pathological bone preparations. Stemming from this, the preparations in this chapter have made it clear that previous simplifications of the focus areas of the seventeenth- and eighteenth-century Leiden anatomists, such as 'the curiosities from the old Anatomical Theatre', 'elegant Albinus', and 'Sandifort and his pathologies', give a very fragmented impression of the anatomical collections. Certain types of preparations were made and collected more or less continuously throughout the entire eighteenth century, yet their meanings could change over time, as is vividly illustrated by the case of the wet bone preparations. The Leiden anatomists were skilled at re-interpreting and re-using already existent preparations, each one using his own bodily experiences with a certain type of preparation, and his own sense of elegance and perfection, in doing so. This would make an attempt to write a history of the skeleton in the eighteenth century an almost impossible tour de force, due to its frequent fragmentation, and entanglement with other body parts, and due to the sheer scope it opens up of endless meanings and uses.

Finally, this chapter has also made clear two important things about aesthesis. One is that the aesthesis of anatomy in eighteenth-century Leiden, although rooted in seventeenth-century anatomical practice with its strong focus on sensory perception and experiment, distinguished itself from the epistemic culture of that period by the quest for, and questions about, elegance and beauty that shaped the eighteenth-century preparations, and by the less obvious presence of moral lessons. The other is that although certain aspects of aesthesis were still influencing the Leiden anatomists and their creating and collecting of preparations by the late eighteenth century, a shift was discernible from a deeply-felt need to prove one's own elegance and perfection as an anatomist in elaborately manufactured preparations that stressed both the perfection of the Maker and the anatomist, to an attitude that focused more strongly on the collecting of occasionally elegant and perfect preparations that could also have been preserved by others. This fits in with the changes in the character of the disciplines of anatomy and physiology that took place by the end of the eighteenth century, previously described by Cunningham. Other factors played a role as well; on the one hand there was the increasingly commonly available knowledge about preparation techniques, and the greater numbers of lasting preparations stemming from the circulation of that knowledge. On the other, new ideas about anatomy research and teaching meant greater numbers of preparations were needed, and these simply could not be assembled through the dissections done by one or two professors.

Aesthesis in eighteenth-century Leiden anatomy—the combined factors of a search for beauty and elegance, the reliance on sensory perception, a wish to deal with disgust, and the need for commodification of the human body—changed in the course of the century, as did the anatomical preparations that were the material sediments of that epistemic culture. Sometimes one aspect was more prominently present in the materiality of anatomical preparations than the others, and the ideas about the role of the most important factors—sensory perception and seeking elegance—varied and even eventually started waning, but the variously leading aspects of aesthesis had not disappeared by 1800. In fact, as Hieke Huistra points out in her book on the Leiden anatomical collections in the nineteenth century, anatomy by then still had an 'aesthetic side', and dealing with disgust and objectification of the body were unabatedly important aspects of anatomy.[66] Moreover, the discipline of anatomy changed fundamentally in the first decades of the nineteenth century.[67] Because of the changes in the extent and manner in which aspects such as seeking perfection and dealing with disgust defined anatomy, aesthesis does not work as an explanatory concept for the epistemic culture of anatomy and most of the resulting collections in the nineteenth century. The changes in how the Leiden bone preparations collections were dealt with in the last decades of the eighteenth century thus foreshadows the end of the specific epistemic culture of aesthesis in anatomy.

66 Huistra, "Preparations on the Move."
67 Cunningham, "The Pen and the Sword."

CHAPTER 8

Conclusion: Aesthesis and the Future of Historical Anatomical Collections

The combinations of aesthesis as a historiographical analytical category with the materiality of the eighteenth-century Leiden anatomical collections as a starting point has enabled a new understanding of both the particular epistemic culture of the eighteenth-century Leiden anatomical collections and specific kinds of eighteenth-century anatomical preparations. Until now, quests for beauty and perfection in epistemic cultures were almost always described as either independent side effect or subordinate to the process of gaining knowledge. Yet the anatomical preparations discussed in this book show that the search for beauty and perfection can be essential to and inextricably intertwined with gaining knowledge. The concept of aesthesis overcomes this common historiographical problem in describing and understanding epistemic cultures. Moreover, aesthesis distinguishes itself from the confused and often misused term aesthetic(s) by returning to the essence of Alexander Baumgarten's eighteenth-century *Aesthetica*: the science of things known through sensory perceptions, which is inextricably connected with a sense of beauty. This combination of sensory perception and a sense of beauty necessarily also includes the development of strategies to deal with the visceral disgust encountered in the processes of gaining anatomical knowledge.

Aesthesis differs from the early modern cultures of wonder and curiosity described by Daston and Park as it does not allow for the static display of uncontested single miracles—it relies heavily on bodily involvement with research materials. Moreover, aesthesis is largely tacit, not so much an individual attitude as an unnamed property of an epistemic culture. Because of these aspects, the materiality of the preparations created and used by the Leiden anatomists can be best understood through the concept of aesthesis. It is in this materiality that aesthesis substantiates. As the creating of preparations of human anatomy involves the rendering stable of otherwise perishable body tissue, and because anatomical preparations were frequently traded and exchanged, aesthesis also includes aspects of objectification, domestification and commodification. Because aesthesis enables an understanding of an epistemic culture accessible through objects, it transcends traditional historiographies such as biographies, institutional histories and colonial history.

 | DOI 10.1163/9789004262775_009

Therefore it provides us with a new, integrated understanding of eighteenth-century Leiden anatomical preparations that were previously often seen as unrelated objects.

Although it is difficult to pinpoint when aesthesis first emerged in Leiden anatomy, an analysis of the materiality of eighteenth-century mercury-injected preparations of the lymphatic system suggested that aesthesis is rooted in seventeenth-century experimentalism and chemical and anatomical practices. At the start of the eighteenth century mercury was associated with qualities like penetration, cleansing, and resurrection. Mercury and sulphur were the two basic elements of transmutational chemistry, the two opposites that could merge into one, wet and cold versus dry and hot, the queen and the king. The inherent redness of blood was a novelty, and in the Hippocratic humoral theory that was still predominant in the seventeenth century, blood was the hot, dry nourishment concoted in the liver to feed the organs. The mysterious lymphatic system, with its almost invisible fluids appeared to have to do something with the arterial system, yet was at the same time clearly distinct. What better way to visualize blood and lymphatic vessels than with hardening masses of red sulphur and silvery mercury? Its attractive silvery shimmer and its penetrating, resurrecting yet elusive character, made mercury the perfect material for anatomists to make visible a largely unknown, hard to discern bodily structure and to simultaneously prove their skills in the newly developed techniques of preparing and injecting hardening fluids into body vessels.

Although some handbooks on preparation instruments and techniques were available, these were far too general and unspecific to follow them like strict manuals. Becoming an anatomist meant not just learning the anatomy of the human body by heart from skeletons and anatomical atlases created by others, but acquiring skills of dissecting and preparing through endless hands-on practice, through trial and error. By the mid-eighteenth century, a skilled anatomist not only knew all the details of human anatomy, he could also display them to others in corpses and preparations; he had developed his own instruments, injection masses, and preservation methods. He stoically dealt with the gore and mess of dissection and decaying bodies in order to understand and display the perfect beauty of human anatomy. For young anatomists such as Eduard Sandifort, whose careers peaked in the second half of the century, materials like mercury had lost their transmutational associations. Yet when the research of the lymphatic system intensified again in from the 1760s, mercury was still their material of choice for exploring it. After all, mercury remained a superbly penetrating, versatile substance, and because it is so hard to handle, an anatomist able to make a good mercury preparation was considered an accomplished anatomist well into the nineteenth century.

From the materiality of the mercury-injected preparations, it becomes clear that refined practical skill and a reliance on one's own sensory perceptions were constant factors throughout the eighteenth century in Leiden anatomy, even if the meanings of certain materials were subject to change.

The importance of elegance in eighteenth-century Leiden anatomy can be seen in all preparations, but shows most vividly in the work of Bernard Siegfried Albinus, who was listed as the first of the 'elegant anatomists' by Albrecht von Haller in his 1774–1777 historiography of anatomists. Albinus certainly strove for elegance, in the dual eighteenth-century sense of the word: he wanted to demonstrate his refined taste and his precision and perfection in his skills and in the anatomical preparations he created. However, as emerged from the analysis of mercury-injected preparations, a strong sense of elegance already played a part in the anatomical practices of Albinus' predecessors like Antony Nuck. The internal evidence of the materiality of preparations of limbs combined with female genitals and sensory organs by Frederik Ruysch and Albinus respectively shows that the latter was inspired the former's craftsmanship and hermeneutics, and used them to communicate his own findings and theories. In the early decades of the eighteenth century, the art of creating lasting wet anatomical preparations was still in its infancy. Human tissue was frequently combined with materials such as textile, lace, red wax and plants to direct the gaze of the observer and to invest preparations with both anatomical and allegorical meanings.

Creating anatomical preparations thus offered Albinus endless opportunities to investigate the functioning of the senses and the principles of life, while simultaneously creating preparations that served both as educational tools and proofs of his own perfect skills and sense of beauty. This multiplicity is best illustrated in two of his preparations. The first, a child's hand holding an eyelid with choroid membrane dangling from a string looks curiously natural at first sight, except for the lace-rimmed sleeve. The sleeve turns out to be a way of directing the gaze of the observer, literally covering up the reminder of the separation from the rest of the body; its lace rim primarily determined by the fashion of the period. Yet on closer inspection it turns out that the arm too has been thoroughly objectified in the anatomists' investigation into the structure and functioning of the senses: the nails and skin have been removed, the veins injected with red wax. The second preparation is of a glove-like skin of a hand, dangling like a beautiful, refined flower from a sprig of a blossoming plant named *Senecio elegans* by Albinus' contemporary Linnaeus. The preparation was used by Albinus to show that the skin is consists of different layers, and that the sense of touch appears to be situated in the lower layers. At the same time it was a proof of his own refinement and elegance. Now damaged beyond repair, this fragile preparation is also a vivid reminder of why it is important that we study historical anatomical collections while they last—even when well looked after, they have an expiration date.

As both my own experiences in recreating an anatomical preparation and eighteenth-century sources confirm, anatomical practice was necessarily also an exercise in dealing with disgust; the anatomist had to overcome abhorrence with the damaged, putrefying body and apply all his knowledge and skill to transform it into elegant preparations that showed the ideal human body. For a long time the focus of anatomical preparations was on the ideal, healthy body, but in the course of the eighteenth century the diseased and deformed body became increasingly important for the anatomist's work. This meant that previously developed strategies to make preparations less disgusting, like the covering up of stumps with sleeves, were no longer sufficient. Eighteenth-century accounts of disgusting anatomical experiences do not support the supposition that we are simply queasier than our eighteenth-century counterparts. However, anatomists seeking beauty and perfection were keen on regulating the repulsive sensory aspects of anatomy and deployed various strategies for dealing with this disgust.

In the case of smallpox, this becomes very clear from a preparation by Albinus of an ear with a tiny smallpox mark. The apparent acquiescence with which the perfection-obsessed Albinus accepted this blemish on a preparation that was essentially about the sense of hearing opens up to the omnipresence of the ravages of smallpox in eighteenth-century medicine and society. Lay people and anatomists alike regulated the anguish caused by smallpox, its lasting marks and the topical debate on inoculation through poetry. Yet literary strategies could not solve all disgust in anatomical practice, as appears from the case of a dog with a cleft palate from the Leiden anatomical collections. However horrifying severe congenital deformities like cleft palate and anencephaly were, eighteenth-century anatomists like Van Doeveren and Sandifort researched them to find out how they originated, and what was their purpose. These were beautiful monsters, in the sense that what is perfect or beautiful is completely fulfilling its purpose; but not necessarily pretty. Van Doeveren's aesthesis, his quest for beauty and perfection in anatomy, led him to interpret monstrosities as yet another proof of nature's astonishing power. The strange combination of revulsion and admiration evoked by eighteenth-century preparations of birth defects can be understood through Korsmeyer's concept of the sublate: they bring home general truths in a particularly vivid manner, and their suggestion of death and decay ensures that the insights they provide are grasped not only with the mind but also with palpable somatic resonance. This book has shown that the experience of disgust upon the confrontation with deformity and ugliness in human anatomy is a form of core disgust that has not changed substantially since the eighteenth century, and aesthesis was as much about dealing with these negatives of the sensory-informed acquisition of knowledge as it was about its positives of beauty and perfection—actually, the two opposites are inextricably connected.

Commodification of (parts of) the dead human body, rendering it stable and tradable, and connected to that its objectification and domestication, was an aspect of eighteenth-century aesthesis in anatomy that becomes most vividly clear in preparations shipped back from colonies to the dominant centre of knowledge. The materiality of ten preparations of foetuses and newborns of supposedly African and Asian descent from the long eighteenth century in the Leiden anatomical collections forms an unexpected window on how the aspects of aesthesis did not only influence anatomists in Leiden, but also medical men who were apparently familiar with Leiden anatomy but who travelled to the colonies. Although a lack of sources makes it impossible to define exactly who made these preparations, their materiality is a very rich source of information about their possible origins. Decorations of strings of coloured beads around necks, waists and limbs unearth knowledge of tribal traditions and indigenous medical knowledge on the eighteenth-century West African coast. Notwithstanding the atrocities of the slave trade in this era and area, the preparations of 'African' babies are not simply the commodification of the exotic other, but also the material remains of a profound interest in human variety and the knowledge and habits of other peoples.

Something similar is the case with two preparations of 'Asian' babies, probably from the Dutch East Indies. These preparations have also been decorated with bead strings, this time of yellow beads that probably symbolize an apparently long forgotten treasure of the island Timor: carnelian beads. The 1778 VOC weapon penny now at the back of the necklace was arguably originally on the chest of the foetus, like the silver snuff boxes worn as penchants on carnelian bead necklaces by affluent Timorese in the eighteenth century. A rattle-like berry in the hand of one of the babies appears to be nutmeg, a commodity highly valued by European colonialists and the indigenous population alike. The many meanings and uses of nutmeg, medical, political and social, combined with the yellow beads and the VOC 'penchant' and the fact that these preparations were made most likely in a period when the Dutch were quickly loosing power in the East Indies, make these preparations tremendously rich objects with a variety of meanings. The maker of these preparations was undoubtedly a man with medical training, who purposefully combined a human foetus with these decorations to show his anatomical and medical skills, his sense of elegance and beauty and his cultural and political awareness so he could stabilize and ship these bizarre yet endearing babies to what to him must have been the dominant centre of knowledge, of aesthesis.

The end of aesthesis in Leiden anatomy is as hard to determine as its emergence, and like most epistemic cultures it only disappeared gradually, not overnight. Tracing changing uses, preparations and meanings of bones in the eighteenth-century Leiden anatomical collections provides us with insights about the end of aesthesis. From the disappeared materiality of the skeleton with which Albinus literally wrestled for the creation of his anatomical atlas, we learn that this period, the first half of the eighteenth century, was the absolute heyday of aesthesis in anatomy. The wet bone preparations made by Wouter van Doeveren show that aesthesis was still influential in the second half of the eighteenth century, with anatomists gauging their own skills and elegance to those of their predecessors and teachers. However, in this period, the importance of the aspects of aesthesis started to change, in concurrence with the disciplinary shift in anatomy and physiology described by Cunningham.

The traditionally theoretical discipline of physiology became increasingly experimental, and Albinus' work on osteogenesis led his successors to new research programs on and understandings of bone diseases. Hence bone pathology became an important part of hands-on anatomy and physiology teaching at Leiden University. This meant that the small Leiden collection of pathological bone preparations needed to be expanded quickly. As preparation techniques and skills were now more widely spread, it became less important for anatomists that each and every piece in their collection was a result of their own sensory involvement and a proof of their own elegant anatomical skills. Moreover, this meant that pathological preparations and collections were increasingly popular commodities, and that old preparations were invested with new uses and meanings. Dealing with disgust remained an important part of the epistemic culture of anatomy, but the accent shifted from concealing the disgusting in preparations to investments in practical facilities such as a well ventilated and heated preparation room in the basement of the university's anatomical theatre to decrease nasty smells.

The previous shows that aesthesis in anatomy was disappearing by 1800; the importance of its various aspects for the discipline were changing. Some of them, like the search for and display of beauty and perfection in anatomy and anatomical preparations, were waning, while others, like commodification and exchange, became more important. Reliance on one's own sensory perceptions and experiments remained important, but was increasingly complemented with the use of preparations and handbooks created by others. Aesthesis thus becomes insufficient as an analytic concept for nineteenth-century anatomy; aesthesis was a particular epistemic culture of the long eighteenth century.

By using aesthesis as an analytical category and the materiality of anatomical preparations to understand the epistemic culture of which the eighteenth-century Leiden anatomical collections are the result, I have put the actual objects that constitute these collections centre-stage without losing sight of the actors, work and social structures from which they emerged, thus transcending traditional historiographical categories. The concept of aesthesis can also be applied in future research on eighteenth-century epistemic cultures, not necessarily anatomical ones, to gain a better understanding of the role of ideals of beauty and perfection, sensory perception, dealing with disgust and commodification in processes of gaining knowledge. In the case of the eighteenth-century Leiden anatomical collections, aesthesis helps to understand the sometimes seemingly paradoxical elements that played such an important role in the creation and initial use of these preparations: life and death, a quest for beauty and perfection and dealing with the disgusting corporeality of anatomical practice, gaining knowledge through relying on one's own senses and the directing of the senses of others, creating preparations to prove one's skills and to commodify and explore the exotic other.

Although this book is not primarily about the display of anatomical collections, the new understanding it provides of the eighteenth-century Leiden collections may hold some implications for the preservation and display of both these and other historical anatomical collections. It is desirable and understandable that the keepers of historical anatomical collections want to thread carefully when it comes to publicly displaying human remains—even if they are over three centuries old and anonymous. However, as the medical collections curator and scholar Ken Arnold has pointed out, it is very well possible to display historical anatomical collections without creating a freak show, and the sensational aspects of medical history actually can be a key to the special significance of the subject. As this book has once again made clear, preparations like these are as much cultural as medical heritage, and they convey complex information about past epistemic cultures and social conventions. Moreover, it is apparent that they can only be properly understood by the modern visitor when additional information about their origins and materiality is provided. Hence it is expedient to display them, preferably not as singularities but as the material results of a bigger culture, in a contextualized public environment like a science museum.

The fragility of many historical anatomical preparations means that displaying them may cause conflict with preservation issues, and the sheer magnitude of some historical collections may mean displaying them is just not an economically feasible option. However, modern technology—either used independently or within the museum space—can forestall these problems. Wonderful examples can be found for example on the website of the Kunstkamera, the website of the Groningen University anatomical collection, at the London Hunterian Museum, and at the Utrecht University Museum. On the Kunstkamera website, the remaining preparations of Frederik Ruysch can be searched and viewed from different angles, and descriptions are linked to preparations wherever possible. When the old Groningen University anatomical collections was no longer used for teaching and donated to the Groningen University Museum, it was too big to integrally put it on display at the museum. Therefore curators created a website that is a virtual tour of the old anatomical collection, in its original cabinets. Each preparation is clickable, so a magnified image can be seen together with a description. By doing this, the original display space is virtually preserved and the entire collection remains accessible. At the London Hunterian Museum, a substantial part of the eighteenth-century Hunter collection is on public display, and almost all preparations can be found in an online database too, provided with images, descriptions, and source references. At the Utrecht University Museum, space and preservation considerations have led curators to select the early nineteenth century anatomical cabinet of professor Jan Bleuland for almost integral display. As the original cabinets are unsuitable for large explanatory plaques, these have been replaced with touch screens in front of the cabinet that allow visitors to access additional information on each preparation.

The eighteenth-century Leiden anatomical collections are at least as rich as the examples mentioned above, and will hopefully become more accessible in the near future. In terms of cost and accessibility, both initially and in the long run, a digital exhibition would be much cheaper than a physical display. Preservation issues are also likely to be easier to manage this way. The Leiden collections are very well maintained and managed now, and this book has shown how important it is to preserve this kind of material culture—after all, materiality is a unique source to increase our understanding of past epistemic cultures. Displaying the Leiden collections both physically and digitally increases their chances of survival. The additional benefit of displaying the collections digitally would be that when some of these fragile preparations eventually perish (and some of them inevitably will), although a small part the materiality of the collections is lost, the understanding of the remaining preparations will benefit from the documentation on what is no longer there.

However, whatever the gains of the digital age may be for collections like these, digitalization can never be a replacement for the actual objects themselves. A photograph, no matter how high its quality, can never entirely capture the essence (or agency, or aura) of the actual encounter with the object. In fact, digital access to collections can never be more than a starting point—it actually accentuates the need to preserve and present the materiality of the actual objects. This also becomes clear when we think of other databases of cultural heritage, like those of art museums. It is wonderful that anyone with an internet connection can now look at Picasso's Guernica anywhere in the world—yet no one will argue that because of this there is no longer any reason for the Reina Sofía Museum to keep it on display, or to keep it at all. The experience of studying the painting online is absolutely incomparable to seeing it on display in Madrid. Its size, the structure of the paint on the canvas, the silent awe in which other visitors look at it: none of this experience can be fully communicated through any medium.

The same goes for these preparations. It was standing eye to eye with them (sometimes literally) that made them whisper, that evoked the questions I tried to answer in this book. Scientists, historians, philosophers, curators, artists and ordinary museum visitors from all over the world share this experience, as shows from the numerous books, articles, blogs, and other sources devoted to history of science and medicine collections, and from the often impressive numbers of visitors of such collections. Hence, if the materiality of these preparations is lost, something essential and irreplaceable is lost forever. It is therefore of the utmost importance that we keep looking for ways to preserve these collections and to simultaneously make them accessible and understandable for generations to come. Finally, this book demonstrates the importance of humanities research on historical anatomical collections. As historians and curators, we are the keepers of the cultural heritage that anatomical collections inevitably become when they age. We are in the position to study these objects and communicate their historical context to a wider audience—for example by developing analytical categories like aesthesis and using them to interprete the material remains of epistemic cultures. This kind of research enables new understandings of historical objects, and adds to the on-going discussion on the desirability of maintaining and displaying of human remains in museums.

Bibliography

Manuscript Sources (MS)

Amsterdam University Library Special Collections (UvA)

Albinus, Bernard Siegfried. Letter to Petrus Camper. UB: HSS-mag.: X2: a–b.

Box, W. Dictata in Physiologiam Clarissimi Viri Professoris B.S. Albini, 4 Volumes. Signatuur hs II G 24–27.

Van Doeveren, Wouter. Letter to J. Tak. UB: HSS-mag.: Y 123: o.

Van Doeveren, Wouter. Letter to J. Tak. UB: HSS-mag.: Y 123: k.

Van Doeveren, Wouter. Letter to Salomon de Monchy. UB: HSS-mag.: Y 123: l.

Van Doeveren, Wouter. Letter to Petrus Camper. UB: HSS-mag.: X54.

Van Doeveren, Wouter. Letter to Petrus Camper. UB: HSS-mag.: X54.

Van Doeveren, Wouter. Letter to Wagler. UB: HSS-mag.: IV A 9: 2 [j]

Van Doeveren, Wouter. Letter. UB: OTM: hs. Y 123: o.

Collection Anatomical Museum LUMC, Leiden (AM)

Boddaert, J.E. Inventaris Der Verzameling in Het Anatomisch Kabinet Van De Rijks Universiteit Te Leiden (1892).

Zaaijer, T. Katalogus Der Ras-Schedels, Bekkens En Skeletten in Het Anatomisch Kabinet Der Rijks Universiteit Te Leiden (1893).

Koninklijke Bibliotheek Den Haag (KB; Royal Library The Hague)

Bonn, A. Brief Van Andreas Bonn (1738–1817), Geschreven Aan Johannes Altheer (1758–1840). 121 E 2 10. Autografenverzameling Beeldsnijder van Voshol.

Leiden University Library Special Collections (UL)

AC2/162–180: No. 161; 22 okt / 25 okt., No. 107, 2 November 1822

AC2/228 uitgaand 1822

Tulbagh, Rijk. Letter to Arnout Vosmaer. BPL 246.

Voegen van Engelen, Johan. Letter to Petrus Camper. (1779) BPL 247/26.

Voegen van Engelen, Johan. Letter to Jean Henri Van Swinden. (1784) BPL 755.

Voegen van Engelen, Johan. Two Letters to Petrus Camper. (1781) BPL 885.

Museum Meermanno-Westrianum, The Hague (MMW)

De Sille, J. Kuypers weduwe. Letter to Johan Meerman. 226/67. Helvoirt, 1804.

De Sille, L. Letter to Johan Meerman. 224/87. 's Hertogenbosch, 1802.

Nationaal Archief, The Hague (NA)

Kamer Amsterdam Ingekomen Stukken. Ingekomen Brieven Van Diverse Personen En Instanties Hier Te Lande. Met Bijlage. 1.05.01.02 483. Tweede West-Indische Compagnie (WIC), The Hague.

Stukken Gediend Hebbende Bij De Criminele Processen Op De Kust Van Guinee. Afschriften. 1.05.01.02 151–152. Tweede West-Indische Compagnie (WIC).

Stadsarchief en Athenaeumbibliotheek, Deventer/Enschede (*SA*)

Burman, Petrus. Salutifera Variolarum Insitio. 1776. Printed Poem in bundle of manuscripts.

Zeeuws Archief, Middelburg (*ZA*)

26.1 Koninklijk Zeeuwsch Genootschap der Wetenschappen (KZGW) 02.2.162 J.C. de Man, Naamlijst van voorwerpen van zoölogische aard alsmede van antropologische aard en pathologische voorwerpen (Middelburg 1879), met aanvullingen in handschrift tot, c. 1908.

20.1410 Middelburgse Commercie Compagnie, Journaal van de scheepsdoctor van de ZeeMercuur 1, Manuscript, 1788–1789.

Material First Published before 1800

"Catalogues of the Rarities in the Leiden Anatomy Theatre Kept in the British Library." (1669–1753).

"Catalogus Eener Uitmuntende Verzameling Van Extra Fraaije Medische, Chirurgische, Anatomische, Botanische, Natuurkundige, Geographische, Historische, Reis, Landbeschryvende, Poëtische En Tooneelkundige Boeken. (...)" Middelburg: Keel, Willem Alexander, 1799.

Albinus, B.S. *Academicarum Annotationum Liber I–VIII*. Leiden, 1754–1768.

———. *Dissertatio De Arteris Et Venis Intestinorum Hominis. Adjecta Icon Coloribus Distincta*. Leiden: Theodor Haak, 1736.

———. *Icones Ossium Foutus Humani, Accedit Osteogeniae Brevis Historia*. Leiden, 1737.

———. *Index Supellectilis Anatomicae, Quam Academiae Batavae Quae Leidae Est Legavit J.J. Rau*. Lugd. Bat.: Henridum Mulhovium et Franciscum Schuyl, 1725.

———. *Tabulae Sceleti Et Musculorum Corporis Humani*. Leiden, 1747.

———. *The Explanation of Albinus's Anatomical Figures of the Human Skeleton and Muscles. With an Historical Account of the Work. Translated from the Latin. To Which Is Added, the Explanation of the Supplement to Albinus, Containing a Compleat*

Anatomical Description of the Blood-Vessels and Nerves. And the External Parts of the Human Body. London: John and Paul Knapton, 1754.

Alpinus, Prosper. *Medicina Aegyptiorum: Accessit Huic Editioni Ejusdem Auctoris Liber De Balsamo, Et Rhapontico, Ut Et Jacobi Bontii Medicina Indorum*. Leiden: Gerard Potvliet, 1718.

Anonymous. *De Koeckoecx-Zangh Van De Nachtuylen Van Het Collegie Nil Volentibus Arduum Huylende Met Eenen Naare Geest: Dr. Ruysch En Boeckelman Zijn De Beest. Of Een Blyeyndende Zamenspraak Tusschen Een Doctor, Proponent, En Poeet*. 1677.

Anonymous. *Holland: A Jaunt to the Principal Places in That Country*. London: W. Hay, 1775.

Azzoguidi, G., J.B. Paletta and J. Brugnoni. *Anatomische Schriften; herausgeg. von E. Sandifort; Aus dem Latein*. Heinr. Tabor, 1791.

Barrow, John. *Dictionarium Medicum Universale: Or, a New Medicinal Dictionary. Containing an Explanation of All the Terms Used in Physic, ... Botany, &C.* .... 2 vols. London, 1749.

Barry, James, John Opie, and Henry Fuseli. *Lectures on Painting by the Royal Academicians*. Elibron Classics. Ed. Wornum, Ralph N. Replica of 1848 edition by Henry G. Bohn, London (fascimile reprint) ed. London Adamant Media Corporation, 2001.

Baumgarten, Alexander. *Reflections on Poetry*. Trans. Holther, William B. Berkely. University Of California Press, 1954.

———. *Aesthetica. Philosophische Bibliothek. lateinisch-deutsch; übers., mit einer Einführung, Anmerkungen und Registern*. 2 vols. Hamburg: Meiner, 2007.

Bilderdijk, W. "Antwoord Op De Vraag Van De Maatschappy Der Nederlandsche Letterkunde Te Leyden, Hebben De Dichtkunst En Welsprekendheid Verband Met De Wijsbegeerte; En Welk Nut Brengt Dezelve Aan De Eene En Andere Toe?" in *Werken van de Maetschappy der Nederlandsche Letterkunde te Leyden*. vol. 6, 1783, 1–200.

Blankaart, Stephan. *Anatomia Reformata, Sive Concinna Corporis Humani Dissectio, Ad Neotericorum Mentem Adornata, Plurimisque Tabulis Chalcographicis Illustrata. Accedit Ejusdem Authoris De Balsamatione, Nova Methodus a Nemine Antehac Hoc Modo Descripta*. Lugduni Batavorum: Cornelium Boutestein, Jordanum Luchtmans, 1688.

———. *Anatomia Reformata, Sive Concinna Corporis Humani Dissectio, Ad Neotericorum Mentem Adornata Editio novissima plurimis recens inventis, tabulisque novis emendatior ac locupletior. Accedit ejusdem authoris De balsamatione nova methodus, à nemine antehac similiter descripta*. Leiden: Boutestein & Luchtmans, 1695.

———. *Nieuw Lichtende Praktyk Der Medicynen: Gefondeert Op De Gronden Van De Deftighste Autheuren Deses Tijdts: Nevens De Hedendaagse Chymia, Als Ook De Nederlantsche Apothekers Winkel; Rijkelijk Met Inlantsche Genees-Middelen Voorsien.* Amsterdam: Jan Claesz. ten Hoorn, 1678.

———. *Nieuwe Konst-Kamer Der Chirurgie of Heelkonst. Neffens Een Korte Ontleeding Van De Pokken En Alle Desselfs Toevallen. 5th ed.* Amsterdam: Jac. ten Hoorn, 1680.

———. *Nieuw-Ligtende Praktyk Der Medicynen.: Nevens Eene Verhandeling Van De Hedendaagze Chymie. De 7e Dr. Merkelijk Verbetert En Nagezien.* Rotterdam: Hermanus Kentlink, 1735.

Boerhaave, Herman. *De Mercurio Experimenta.* Utrecht: Jurriaan van Paddenburg, 1735.

———. *Praelectiones Academicae in Proprias Institutiones Rei Medicae. Edidit, Et Notas Addidit Albertus Haller. Vol. 1. 6 vols.* Amsterdam: J. Wetstenium, 1747.

———. *Sermo Academicus De Chemia Suos Errores Expurgante.* Leiden: Petrus van der Aa, 1718.

———. *Some Experiments Concerning Mercury. Translated from the Latin, Communicated by the Author to the Royal Society.* London: J. Roberts, 1734.

———. E. Kegel-Brinkgreve, and Antonie M. Luyendijk-Elshout. *Boerhaave's Orations.* Leiden: Sir Thomas Browne Instituut, 1983.

Bonn, Andreas. *Redevoeringen Ter Inwijding Der Volbouwde Tekenzaal... En Van De Gehoorzaal... Der Natuurkunde, in Het Gebouw Der Maatsch. Felix Meritis.* Amsterdam, 1790.

Bosman, Willem. *Nauwkeurige Beschrijving Van De Guinese Goud- Tand- En Slavenkust.* Amsterdam: Isaak Stokmans, 1709.

Burke, Edmund. *A Philosophical Enquiry into the Origin of Our Ideas of the Sublime and the Beautiful.* Oxford: Oxford University Press, 1990.

Camper, Petrus. "Dissertatio Optica De Visu." Dissertation. Leiden University, 1746.

———. *Verhandeling Van Petrus Camper, over Het Natuurlijk Verschil Der Wezenstrekken in Menschen Van Onderscheiden Landaart En Ouderdom; over Het Schoon in Antyke Beelden En Gesneedene Steenen. Gevolgd Door Een Voorstel Van Eene Nieuwe Manier Om Hoofden Van Allerleye Menschen Met Zekerheid Te Tekenen. Naa Des Schrijvers Dood Uitgegeeven Door Zijnen Zoon Adriaan Gilles Camper.* Utrecht: B. Wild en J. Altheer, 1791.

Chais, Charles. "Essaie Apologétique Etc." *Verhandelingen uitgegeven door de Hollandse Maatschappy der Weetenschappen te Haarlem* 1 (1754): 600–60.

Coopmans, Gadso. *Oratio De Medicamentis Indigenis Ad Morbos Belgio Familiares Feliciter Depellendos Suffecturis.* Franeker: Willem Coulon, 1774.

———. *Varis; Sive Carmen De Variolis.* Franeker: Wilhelm Coulon, 1783.

Curtis, William. *The Botanical Magazine, or Flower-Garden Displayed.* Vol. 7. London: Stephen Couchman, 1792.

Dapper, Olfert. *Naukeurige Beschrijvinge Der Afrikaensche Gewesten Van Egypten, Barbaryen, Libyen, Biledulgerid, Etc.* Amsterdam: Jacob van Meurs, 1668.

De Bils, Louis. *Vertooch Van Verscheyde Eyghene Anatomische Stucken.* Amsterdam: Nicolaes van Ravesteyn, 1655.

De Graaf, Regnerus. *De succo pancreatico: or, a physical and anatomical treatise of the nature and office of the pancreatick juice.* London, 1676.

———. *Korte Beschryving Van 'T Gebruyk Der Spuyt in D' Ontleedt-Konst. Met Een Inleiding Door Dr. H.L. Houtzager.* Delft: Reinier de Graaf Stichting, 1989.

Engelhard, Nicolaas. *Feriae Aestivae Groninganae Exhibentes Variorum Variiargument Dissertationum Praecipue Philosophicarum*, Vol. 2. Groningen: Jacob Sipkes, 1736.

Erndl, Christian Heinrich. *The Relation of a Journey into England and Holland, in the Years, 1706, and 1707. By a Saxon Physician, in a Letter to His Friend at Dresden, Wherein Are Contain'd Many Remarkable Passages and Curious Observations in Anatomy, Surgery, Physick, and Philosophy.* London: John Morphew, 1711.

Gaubius, H.D. "Inaugurele Rede, Waarin Wordt Aangetoond, Dat De Scheikunde Met Recht Een Plaats Verdient Onder De Akademische Wetenschappen. (Vertaling Van "Oratio Inauguralis Qua Ostenditur Chemiam Artibus Academicis Jure Esse Inserendam," 21 Mei 1731)." *Opuscula Selecta Neerlandicorum De Arte Medica.* Ed. Rossen, F. van. Vol. 1. Amsterdam, 1907: 171–228.

Graeuwen, Paulus. *De Anatomiae Pathologicae Utilitate Ac Necessitate.* Groningae: Hajo Spandaw, 1771.

Hemsterhuis, Frans. 1771, 1776: see Hemsterhuis 2001.

Houttuyn, Martinus. *Natuurlyke Historie of Uitvoerige Beschryving Der Dieren, Planten En Mineraalen, Volgens Het Samenstel Van Den Heer Linnæus. Met Naauwkeurige Afbeeldingen, Eerste Deels, Tiende Stuk. De Kruiden.* Amsterdam: Bij de erven F. Houttuyn, 1766.

Hume, David. *An Enquiry Concerning Human Understanding.* Mineola NY: Dover Publications, 2004.

Kaau, Abraham. *Declamatio Academia De Gaudiis Alchemistarum.* Leiden: Samuel Luchtmans, 1737.

Kant, Immanuel. *Kritik Der Reinen Vernunft. Kritik Der Praktische Vernunft Kritik Der Urteilskraft. Ungekürtzte Sonderausgabe Zum Kantjahr 2004.* Wiesbaden: Fourier Verlag, 2003.

La Mettrie, Julien Offray de. *L'homme Machine.* Leyde: Elie Luzac, fils, 1748.

Lindeboom, G.A. (ed) *Observationes anatomicæ Collegii privati Amstelodamensis. Pars prior (1667) et altera (1673). Fascimile with an introduction by G.A. Lindeboom.* Nieuwkoop: B. De Graaf, 1975.

Lyser, Michael. *Culter Anatomicus, Hoc Est: Methodus Brevis Humana Incidendi Cadavera.* Hafniae, 1665.

———. *The Art of Dissecting the Human Body: In a Plain, Easy, and Compendious Method. With the Several Instruments Necessary, Curiously Engraved;... Translated from the Latin of Lyserus. Trans. G. Thomson, M.D.* London: Joseph Davidson, 1740.

Mascagni, Paolo. *Prodrome D'un Ouvrage Sur Le Systè me Des Vaisseaux Lymphatiques.* Sienne, 1784.

———. *Vasorum Lymphaticorum Corporis Humani Historia Et Ichnographia.* Senis, 1787.

Mendelssohn, Moses. *Briefe, Die Neueste Litteratur Betreffend, 4. Januar 1759–4. Juli 1765. Ed. bearb. von Eva J. Engel; mit Beiträgen von Michael Albrecht, Elisabeth Blakert und Hans Lausch.* 2 vols. Stuttgart-Bad Cannstatt: Frommann-Holzboog, 2004.

———. and Gerrit Brender à Brandis. *Wijsgeerige Verhandelingen, Brieven En Gesprekken, / Van Moses Mendelszoon.; Uit Het Hoogduitsche Vertaald, En Met Aantekeningen, En Het Leven Van Den Autheur, Verrijkt, Door G. Brender a Brandis.* Leiden: Pieter Pluygers, 1788.

Monro, Alexander, and Johann Friedrich Meckel. *Opuscula Anatomica De Vasis Lymphaticis. 2 vols.* Leiden: Lipsius, 1760.

Nesbitt, Robert. *Human Osteogeny Explained in Two Lectures, Read in the Anatomical Theatre of the Surgeons of London.* London: T. Wood, 1736.

Neumann, Caspar. *Grondlyke En Met Proeven Bewesene Medicinale En Natuurkundige Chymie.* Leeuwarden: H.A. de Chalmot, 1766.

Noortwijk, Wilhelmus. *Uteri Humani Gravidi Anatome Et Historia.* Lugduni Batavorum: Joan. et Herm. Verbeek, 1743.

Nuck, Antony. *Adenographia Et Uteri Anatome Nova.* Leiden: Luchtmans, 1691.

———. *De Ductu Salivali Novo, Saliva, Ductibus Oculorum Aquosis Et Humore Oculi Aqueo.* Leiden: Pieter van der Aa, 1685.

———. *Genees- Ofte Heelkundige Handwerken En Ondervindingen. Met Plaetverbeeldingen: Uit Het Latyn Vertaelt, En Met Een Voorreden Verrijkt Door Regnerus Van Overschuur.* Leiden: Samuel Luchtmans, 1740.

———. *Operationes Et Experimenta Chirurgica.* Leiden, 1692.

———. *Operationes Et Experimenta Chirurgica.* Leiden: Samuel Luchtmans, 1733.

Paterson, Samuel. *Museum Falconarium. A Catalogue of the Entire and Capital Museum of Anatomical Preparations and Other Subjects of Natural History, a Great Variety of Chirurgical, Anatomical, and Philosophical Instruments; Medicaments, Cabinets, Preparation-Glasses, and Other Effects; of the Late Mr. Magnus Falconar, etc.* London, 1778.

Pole, Thomas. *The Anatomical Instructor; or an Illustration of the Modern and Most Approved Metods of Preparing and Preserving the Different Parts of the Human Body and of Quadrupeds by Injection, Corrosion, Maceration, Distention, Articulation, Modelling, &C.* London: Couchman & Fry, 1790.

Poole, R. *A Journey from London to France and Holland: Or, the Traveller's Useful Vade Mecum Vol. 2. 2 vols.* London: Duncomb, 1742.

Quincy, John. "Lexicon Physico-Medicum: Or, a New Physical Medicinal Dictionary, Explaining the Difficult Terms Used in the Several Branches of the Profession, and in Such Parts of Natural Philosophy as Are Introductory Thereunto.... With New Improvements from the Latest Chymical and Mechanical Authors." 5th ed. London: T. Longman, 1736.

Rau, J.J. *De Methodo Anatomen Docendi Et Discendi.* Lugd. Bat., 1713.

Ridley, Humphrey. *The Anatomy of the Brain.* London: Sam. Smith and Benj. Walford, 1695.

Rumphius, Georg Everard. *D'amboinsche Rariteitkamer, Behelzende Eene Beschryvinge Van Allerhande Zoo Weeke Als Harde Schaalvisschen, Te Weeten Raare Krabben, Kreeften, En Diergelyke Zeedieren, Als Mede Allerhande Hoorntjes En Schulpen, Die Men in D'amboinsche Zee Vindt: Daar Beneven Zommige Mineraalen, Gesteenten, En Soorten Van Aarde, Die in D' Amboinsche, En Zommige Omleggende Eilanden Gevonden Worden.: Verdeelt in Drie Boeken ... Bevat Ook Met Koptitel: Beschryving En Verdeeling Der Amboinsche Hoornen En Schulpen, Door Den Heer Heer Sipman. Vol. 1–3.* Amsterdam: François Halma, 1705.

———. *Het Amboinsche Kruid-Boek. Dat Is, Beschryving Van De Meest Bekende Boomen, Heesters, Kruiden, Land- En Water-Planten, Die Men in Amboina, En De Omleggende Eylanden Vind.* Vol. 1–6. Amsterdam: François Changuion, Jan Catuffe, Hermanus Uytwerf, 1741–1750.

Ruysch, Frederik. *Alle de ontleed- genees- en heelkundige werken van Frederik Ruysch.* Amsterdam: Janssoons van Waesberge, 1744.

———. *Effigies Penis Humani, Injecta Cera Praeparati Exhibens Inventa Anatomica Aliquot Nova Et Proprio Colore Typis Impressa a Joanne Ladmiral.* Leiden: Cornelis Haak, 1790.

———. *Dilucidatio Valvularum in Vasis Lymphaticis Et Lacteis.* Den Haag: Harmani Gael, 1665.

———. *Ontleedkundige Verhandelingen over De Vinding Van Een Spier in De Grond Des Baar-Moeders: Te Voren Noyt Ontdekt. Mitsgaders Een Nader Onderwys Wegens De Verlossing Der Nageboorte Der Barende Vrouwen.* Amsterdam, 1725.

———. "Thesauri Anatomica." 13 vols. Amsterdam: UvA, 1701–1728.

Sandifort, Eduard. *Eduardi Sandifort Oratio De Circumspecto Cadaverum Examine Optimo Practic Medicin Adminiculo.* Leiden: Sam. et Joh. Luchtmans, 1772.

———. *Observationes Anatomico-Pathologicae. 4 vols.* Leiden: P. v.d. Eyk et D. Vygh, 1777–1781.

———. *Natuur- En Genees-Kundige Bibliotheek, Bevattende Den Zaakelyken Inhoud Van Alle Nieuwe Werken, Welke in De Geneeskunde En Natuurlyke Historie, Buiten Ons Vaderland Uitkomen. 11 vols.* 's-Gravenhage: Pieter van Cleef, 1765–1775.

———. and Gerard Sandifort. *Museum Anatomicum Academiae Lugduno-Batavae. 4 vols.* Leiden: Luchtmans, 1793–1835.

Schroder, Johannes. *Schatkamer Der Droogbereidende Geneeskonst, En Ontsloten Kabinet Der Simplica.* 3 vols. Mechelen: Lobelia, 1985.

Smith, James Edward. *A Sketch of a Tour on the Continent, in the Years 1786 and 1787, by James Edward Smith, … In Three Volumes. … Vol. 1. 3 vols.* London, 1793.

Terne, Cornelis. "Antwoord Op De Vraag, Voorgesteld Door Het Genootschap Der Kunsten En Wetenschappen Te Batavia, over Het Beste Voedsel Om Nieuw Geboren Kinderen Zonder Borst of Moedermelk Op Te Voeden, Onder De Zinspreuk, Ego Fateor, Me Exeorum Numero Esse Conari, Qui Proficiendo Scribunt Et Scribendo Proficiunt, Aan Den Schryver Van Het Welke De Eerprys Eener Gouden Medaille Is Toegewezen." *Verhandelingen Van Het Bataviaasch Genootschap Der Kunsten En Weetenschappen.* Batavia: Bataviaasch Genootschap der Kunsten en Wetenschappen, 1794.

———. *Dissertatio Medica Inauguralis De Abortu Et Partu Praematuro.* Leiden: Theodor Haak, 1778.

Van Doeveren, W. *Academische Redevoering Van W. Van Doeveren over De Gunstige Gesteldheid Van Groningen Voor De Gezondheid, Afteleiden Uit De Natuurlyke Historie Der Stad…; Nu Uit Het Latyn Vertaald, En Met Eene Verhandelende Voorrede Uitgegeeven Door Matthias Van Geuns.* Groningen: Cornelis Barlinkhof, en Jacob Bolt, 1771.

———. *Bibliotheca Doeveriana, Sive Catalogus Librorum Gualth. Van Doeveren: Quorum Auctio Fiet Publ. 27 Sept. 1784.* Lugduni Batavorum, 1784.

———. *Brief Aan Eduard Sandifort, Behelzende Een Berigt Van Den Gelukkigen Uitslag Der Inentinge Der Kinderpokjes, Onlangs Te Groningen Ondervonden.* 's-Gravenhage, 1770.

———. *Dissertatio De Vermibus Intestinalibus Hominum.* Groningen, 1753.

———. *Musei Doeveriani Catalogus, Contin. Copiam Quadrupedem, Avium…; Quae Collegit Gualth. Van Doeveren: Horum Publ. Auctio Fiet 18 Apr. 1785.* Lugd. Bat., 1785.

———. *Raadgevinge Om De Inentinge Der Ziekte Van 'T Rundvee Op Eene Gemakkelyke, Onkostbare En Voordeelige Manier Te Doen, Uit Eigene Ondervindinge Opgemaakt.* Groningen: Jacob Bolt, 1769.

———. *Sermo Academicus De Recentiorum Inventis Medicinam Hodiernam Veteri Praestantiorum Reddentibus.* Lugd. Bat., 1771.

———. *Sermo Academicus De Sanitatis Groninganorum Praesidiis, Ex Urbis Naturali Historia Derivandis / Dictus Cum Iterum Fasces Academicos Poneret.* Groningae, 1770.

———. *Specimem Observationum Academicarum Ad Monstrorum Historiam, Anatomen, Pathologiam Et Artem Obstetricam, Praecipue Spectantium*. Groningae et Lugd. Bat., 1765.

———. J.H. Voyer, and Daniel Canisius. *Q. D. B. V.! Ex Auctoritate Amplissimorum Goesanae Scholae Curatorum…: Sollemni Discipulorum Promotione, Duabus Duorum Adolescentum Orationibus, Gualtheri Van Doeveren De Inspicienda Amicorvm Fide; Danielis Canisii, Qvibvs Rebvs Rverit Resp. Romana, Inquisituri: Suam Præmissurus Orationem De Vita Academica, Hanc Actionem Ut Cohonestare Frequenti Conventu Benigne Velint, Omnes Literarum Patronos Ac Studiosos Comiter / Invitat J.H. Voyer* Goes: Huisman Jacobszoon, Adriaan, 1747.

———. "Verhandelinge over De Waare Kinderpokjes, Die Meer Dan Eenmaal Denzelfden Mensch Aantasten." *Verhandelingen Van De Hollandse Maatschappye Der Wetenschappen*. Vol. XII. Haarlem: Hollandse Maatschappye der Wetenschappen, 1770: 189–232.

Van Hogendorp, Willem. "Vervolg Der Beschryving Van Het Eiland Timor, Voor Zoo Verre Het Tot Nog Toe Bekend Is, Waar by Gevoegd Is De Beschryving Van Eenige Nabygelegene Eilanden" *Verhandelingen Van Het Bataviaasch Genootschap Der Konsten En Weetenschappen*. Vol. 2. Rotterdam/Amsterdam: Reinier Arrenberg/ Johannes Allart, 1784: 63–101.

Van Rooijen, Adriaan. "Voorrede" in: Philip Miller, *Groot en algemeen kruidkundig, hoveniers, en bloemisten woordenboek, behelzende de manier om moes-, bloem-, vrugt-, kruid-tuinen, wild-bossen, wijngaarden, Oranje-huizen, stook-kassen, enz. aanteleggen, enz. en allerlei gewassen te kweeken… Waarby ook komt… het gebruik der barometers, thermometers, en hygrometers, welke den hoveniers te pas komen*. Leiden: Pieter van der Eyk / De Weduwe Jakob van der Kluis, 1745.

———. *Poëmata*. Leiden: Cornelis van Hoogeveen jr., 1778.

Van Visvliet, E.P. *Catalogus Eener Uitmuntende Verzameling Van Extra Fraaije Medische, Chirurgische, Anatomische, Botanische, Natuurkundige, Geographische, Historische, Reis, Landbeschryvende, Poëtische En Tooneelkundige Boeken*. Middelburg: Keel, Willem Alexander, 1799.

Vesalius, Andreas. *De Corporis Humani Fabrica Libri Septem*. 1543. http://www.nlm.nih.gov/exhibition/historicalanatomies/vesalius_home.html.

Voegen van Engelen, Johan. *De Surinaamsche Artz. Uitgaven Van De Natuurwetenschappelijke Studiekring Voor Suriname En De Antillen. Natuurhistorische Reeks Vol. 109*. Amsterdam: Rodopi, 1981.

Von Goethe, Johann Wolfgang. 1795–6: see Von Goethe 1987.

Von Haller, Albrecht. 1725–1727: see Von Haller 1958.

———. *Bibliotheca Anatomica: Qua Scripta Ad Anatomen Et Physiologiam Facientia a Rerum Initiis Recensentur Vol. 2*. Zurich: Tiguri 1774–1777.

———. and Boerhaave, Herman. *Praelectiones Academicae in Proprias Institutiones Rei Medicae. Edidit, Et Notas Addidit Albertus Haller. Vol. 1.* 6 vols. Amsterdam: J. Wetstenium, 1747.

Von Herder, Johann Gottfried. "Plastik. Einige Wahrnehmungen Über Form Und Gestalt Aus Pygmalions Bildenden Traume." *J.G. Von Herder's Sämmtliche Werke. Zur Schönen Literatur Und Kunst. Vol. 11, Zur römischen Literatur.* Tübingen: Antiquarisch Aufsätze, 1809: 239–363.

Von Uffenbach, Z.C. *Merckwürdige Reisen Durch Nieder-Sachsen, Holland Und England.* 3 vols. Ulm: J.F. Gaum, 1753–1754.

Wagler, Carl Gottfried. *Verzeichnisz Der Anatomischen Präparate / Welche Von Ca. Gottl. Wagler Sind Gesammlet Worden.* Braunschweig, 1779.

Worm, Ole. *Museum Wormianum. Seu Historia rerum rariorum, tam naturalium, quam artificialium, tam domesticarum, quam exoticarum, quae Hafniae Danorum in aedibus authoris servantur.* Leiden: Ludovicum & Danielem Elzevirios, 1655.

Modern Works

Abraham, Lyndy. *A Dictionary of Alchemical imagery.* Cambridge: Cambridge University Press, 1998.

Adams, Percy G. *Travelers and Travel Liars, 1660–1800.* Berkely: University of California Press, 1962.

Alberti, Samuel J.M.M. "Objects and the Museum." Isis 96 4 (2005): 559–71.

———. "Owning and Collecting Natural Objects in Nineteenth-Century Britain." in *From Private to Public: Natural Collections and Museums.* Ed. Berretta, Marco. New York: Science History Publications, 2005: 141–54.

———. *Morbid Curiosities. Medical Museums in Nineteenth-Century Britain.* Oxford: Oxford University Press, 2011.

Alpers, Svetlana. *The Art of Describing: Dutch Art in the Seventeenth Century.* Chicago: University of Chicago Press, 1983.

Ambrose, C.T. "The priority dispute over the function of the lymphatic system and Glisson's ghost (the 18th-century Hunter-Monro Feud)." Cellular Immunology, 2007, Vol. 245 (1): 7–15.

Anthon, Charles. *A Classical Dictionary: Containing the Principle Proper Names Mentioned in Ancient Authors Part Two.* New York: Harper & Brothers, 1842.

Appadurai, Arjun, ed. *The Social Live of Things. Commodities in Cultural Perspective.* New York: Cambridge University Press, 1986.

Aransiola, Joshua Oyeniyi, and Suraj Olalekan Asindemade. "Understanding Cybercrime Perpetrators and the Strategies They Employ in Nigeria." *Cyberpsychology, Behavior, and Social Networking* 14.12 (2011): 759–63.

Arnold, Ken. "Museums and the Making of Medical History." in *Manifesting Medicine*. Ed. Bud, Robert. London: NMSI Trading Ltd, Science Museum, 2004: 145–74.

Ball, Philip. *Elegant Solutions. Ten Beautiful Experiments in Chemistry*. Cambridge: The Royal Society of Chemistry, 2005.

Barasch, Moshe. *Modern Theories of Art: From Winckelmann to Baudelaire, Volume 1*. New York: NYU Press, 1990.

Barge, J.A.J. *De Leidsche Anatomie Herdacht. Toespraak Bij De Opening Van Het Anatomisch Laboratorium Der Rijksuniversiteit Te Leiden Op 2 October 1923 Gehouden*. Leiden: S.C. van Doesburgh, 1923.

———. *Het Geneeskundig Onderwijs Aan De Leidsche Universiteit in De 18de Eeuw*. Leiden: S.C. van Doesburgh, 1934.

———. *De Oudste Inventaris Der Oudste Academische Anatomie in Nederland*. Leiden/Amsterdam: H.E. Stenfert Kroese's Uitgevers Mij N.V., 1934.

———. *De Stichting Van Het Academisch Klinisch Onderwijs Te Leiden Voor 300 Jaren. Herdenkingsrede Gehouden in De Bijeenkomst Van Den Academischen Senaat Der Rijksuniversiteit Te Leiden Op 29 October 1937 Door Den Rector Magnificus*. Leiden: H.E. Stenfert Kroese's Uitgevers-Mij N.V., 1937.

Beardsley, Monroe C. *Aesthetics from Classical Greece to the Present: A Short History*. Tuscaloosa: University of Alabama Press, 1975.

Bertoloni Meli, Domenico. "The Color of Blood: Between Sensory Experience and Epistemic Significance." in *Histories of Scientific Observation*. Eds. Daston, Lorraine and Elizabeth Lunbeck. Chicago and London: The University of Chicago Press, 2011: 117–34.

Bianchi, F. et al. "Environment and Genetics in the Etiology of Cleft Lip and Cleft Palate with Reference to the Role of Folic Acid." *Epidemiologia e Prevenzione* 24.1 (2000): 21–7.

Bots, Hans. "Van Universitaire Gemeenschap Tot Academische Kring. Enige Aspecten Met Betrekking Tot De Opkomst En Ontwikkeling Van Geleerde Academies En Genootschappen in West-Europa." *Genootschap ter bevordering van Natuur-, Genees- en Heelkunde*. Amsterdam, 1976.

Bradt, Hilary. *Slow Devon and Exmoor*. Chalfont St Peter, Bucks: Bradt Travel guides, 2010.

Breman, Joel G., and D.A. Henderson. "Diagnosis and Management of Smallpox." *New England Journal of Medicine* 346 (2002): 1300–08.

Brown, Bill. "Thing Theory." *Critical Inquiry* 28 1 (2001): 1–22.

Brugmans, Sebald Justinus et al. *Pharmacopoea Batava*. Amsterdam: J. Allart, 1805.

Bruijn, Iris. "Ship's Surgeons of the Dutch East India Company in the Eighteenth Century. Commerce and the Progress of Medicine." Leiden University, 2004.

Bucklow, Spike. *The Alchemy of Paint: Art, Science, and Secrets from the Middle Ages*. London/New York: Marion Boyards, 2009.

Bussagli, Marco. *Het menselijk lichaam. Anatomie en symboliek.* Antwerpen: Ludion, 2007.

Carey, Margaret. "Gender in African Beadwork." in *Beads and Bead Makers. Gender, Material Culture and Meaning.* Eds. Scimia, Lidia D. and Joanne B. Eicher. Oxford: Berg, 1998: 83–93.

Chaplin, Simon. "John Hunter and the Anatomy of a Museum." *History Today* 55 2 (2005): 19–25.

Chappel, T.J.H. "The Yoruba Cult of Twins in Historical Perspective." *Africa* 44 (1974): 250–65.

Cheetham, Mark. *Kant, Art and Art History: Moments of Discipline.* Cambridge: Cambridge University Press, 2002.

Clark, Stuart. *Vanities of the Eye. Vision in Early Modern European Culture.* Oxford: Oxford University Press, 2007.

Clarke, Edwin, and C.D. O'Malley. *The Human Brian and Spinal Cord. A Historical Study Illustrated by Writings from Antiquity to the Twentieth Century.* Berkeley/Los Angeles: University of California Press, 1968.

Cole, F.J. "The History of Anatomical Injections." *Studies in the History and Method of Science.* Ed. Singer, Charles. Vol. 2. Oxford: Clarendon Press, 1921: 285–344.

Collingwood, R.G. *The Principles of Art.* Oxford: Clarendon Press, 1938.

Collins, Harry. "Actors' and Analysts' Categories in the Social Analysis of Science." in *Clashes of Knowledge: Orthodoxies and Heterodoxies in Science and Religion.* Eds. Meusburger, Peter, Michael Welker and Edgar Wunder: Springer, 2008: 101–10.

———. *Tacit and Explicit Knowledge.* Chicago: University Of Chicago Press, 2010.

Cook, Harold. *Matters of Exchange: Commerce, Medicine and Science in the Dutch Golden Age.* Yale: Yale University Press, 2007.

Cuhaj, George, and Merna Dudley, eds. *Standard Catalog of World Coins 1701–1800.* 5th ed. Iola WI: Krause Publications Inc, 2010.

Cunningham, Andrew. *The Anatomical Renaissance: The Resurrection of the Anatomical Projects of the Ancients.* Aldershot: Ashgate, 1997.

———. *The Anatomist Anatomis'd. An Experimental Discipline in Enlightenment Europe.* London: Ashgate, 2010.

———. "The Pen and the Sword: Recovering the Disciplinary Identity of Physiology and Anatomy before 1800. Part One. Old Physiology: The Pen." *Studies in History and Philosophy of Biological and Biomedical Sciences* 33 (2002): 631–65.

———. "The Pen and the Sword: Recovering the Disciplinary Identity of Physiology and Anatomy before 1800. Part Two. Old Anatomy: The Sword." *Studies in History and Philosophy of Biological and Biomedical Sciences* 34 (2003): 51–76.

———. "The Principality of Blood. William Harvey, the Blood, and the Early Transfusion Experiments." in Santing, Catrien, and Jetze J. Touber (eds). *Blood-Symbol-Liquid.* Leuven: Peeters, 2012: 193–204.

Dalzel, Archibald. *De Geschiedenis Van Dahomy, Een Binnenlands Koningrijk Van Afrika; Benevens Het Verhaal Eener Binnenlandse Reize in Afrika (...) Naar Het Engelsch Met Plaaten*. Den Haag: J.C. Leeuwenstyn, 1800.

Daniëls, Carel Eduard. *De Kinderpok-Inenting in Nederland: Meerendeels Naar Onuitgegeven Bescheiden Bewerkt: Eene Medisch-Historische Studie*. Amsterdam: Post, 1875.

Dannehl, Karin. "Object Biographies: From Production to Consumption." in *History and Material Culture. A Student's Guide to Approaching Alternative Sources*. Ed. Harvey, Karen. Routledge Guides to Using Historical Sources. London: Routledge, 2009: 123–38.

Daston, Lorraine. "The Empire of Observation, 1600–1800." in *Histories of Scientific Observation*. Eds. Daston, Lorraine and Elizabeth Lunbeck. Chicago and London: The University of Chicago Press, 2011: 81–113.

———. and Peter Galison. *Objectivity*. New York: Zone Books, 2007.

———. and Peter Galison. "The Image of Objectivity." *Representations* 40 (1992): 81–128.

———. ed. *Things That Talk. Object Lessons from Art and Science*. New York: Zone Books, 2008.

———. and Katharine Park. *Wonders and the Order of Nature 1150–1750*. New York: Zone Books, 1998.

De Bolla, Peter. *The Discourse of the Sublime. Readings in History, Aesthetics and the Subject*. Oxford: Basil Blackwell, 1989.

De Clippel, Karolien. "Alchemist / Alghe Mist. Het Beeld Van De Alchemist in De Zeventiende-Eeuwse Schilderkunst Van De Lage Landen." in *De Wijze En De Dwaas. De Alchemist in Woord En Beeld*. Ed. Jacobs, Davy. Brugge: Uitgeverij Van De Wiele, 2007: 27–38.

Den Heijer, Henk. *Naar De Koning Van Dahomey. Het Journaal Van De Gezantschapsreis Van Jacobus Elet Naar Het West-Afrikaanse Koninkrijk Dahomey in 1733*. Zutphen: Walburg Pers, 2000.

———. *De Geschiedenis Van De WIC*. Zutphen: Walburg Pers, 2007.

De Rijcke, Sarah. "Regarding the Brain. Practices of Objectivity in Cerebral Imaging. 17th Century-Present." Dissertation. Groningen University, 2010.

De Rooy, Laurens, and Hans van den Boogaard. *Verzamelaars Van Vorm*. Amsterdam: Vossiuspers UvA, 2009.

De Vries, Ad, and Arthur de Vries. "Rattle." *Elsevier's dictionary of symbols and Imagery*. 2nd ed. Amsterdam: Elsevier, 2004.

Dewhurst, Kevin. *The Quicksilver Doctor. The Live and Times of Thomas Dover, Physician and Adventurer*. Wright: Univeristy of California, 1957.

Dobbs, Betty J.T. *Alchemical Death and Resurrection: the Significance of Alchemy in the Age of Newton*. Washington: Smithsonian Institution Libraries, 1990.

Driessen van het Reve, Jozien J. "Frederik Ruysch in Rusland" *Geschiedenis der geneeskunde* 11 2 (2005): 68–77.

———. and Luuc Kooijmans. *Geloof Alleen Je Eigen Ogen: Frederik Ruysch (1638–1731) Held Van De Nieuwe Wetenschap*. Amsterdam: Universiteitsbibliotheek Amsterdam, 2004.

Drucker, Johanna. *SpecLab: digital aesthetics and projects in speculative computing*. Chicago: University Of Chicago Press, 2009.

Dubin, Lois Sherr. *The Worldwide History of Beads. Ancient, Ethnic, Contemporary*. London: Thames & Hudson, 2009.

Dupré, Sven, and Christoph Lüthy. "Introduction: Silent Messengers. The World of Goods and the Circulation of Knowledge in the Early Modern Netherlands." in *Silent Messengers. The Circulation of Material Objects of Knowledge in the Early Modern Low Countries*. Dupré, Sven, and Christoph Lüthy (eds). Vienna: Lit Verlag, 2011: 1–12.

Eagleton, Terry. *The Ideology of the Aesthetic*. Oxford: Basil Blackwell, 1990.

Eales, Nellie B. "The History of the Lymphatic System, with Special Reference to the Hunter-Monro Controversy." *Journal for the History of Medicine and Allied Sciences* XXIX.3 (1974): 280–94.

Earnshaw, Pat. *Lace in Fashion from the Sixteenth to the Twentieth Centuries*. London: B.T. Batsford Ltd, 1985.

Ebenstein, Joanna. "Morbid Anatomy Blogspot". New York, 2007. 3 May 2012. http://morbidanatomy.blogspot.com/.

Eijgenraam, M.J. "'Menschlievendheid En Eigen Belang'. De Behandeling Van De Slaven Aan Boord Van De Schepen Van De Middelburgsche Commercie Compagnie." *Archief. Mededelingen van het Koninklijk Zeeuwsch Genootschap der wetenschappen* 10 (1990): 77–104.

Elshout, A.M. (also see Luyendijk-Elshout, A.M.) *Het Leidse Kabinet Der Anatomie Uit De Achttiende Eeuw. De Betekenis Van Een Wetenschappelijke Collectie Als Culturhistorisch Monument*. Leiden: Universitaire Pers, 1952.

Fagg, William. *Yoruba Beadwork. Art of Nigeria*. New York: Rizzoli International Publications, 1980.

Fara, Patricia. "Marginalized Practices." The Cambridge History of Science. Ed. Porter, Roy. Vol. 4. Eighteenth-Century Science. Cambridge: Cambridge University Press, 2003: 485–507.

Farram, Steven. "Jacobus Arnoldus Hazaart and the British Interregnum in Netherlands Timor, 1812–1816." Bijdragen tot de taal-, land- en volkenkunde 163 4 (2007): 455–75.

Fontes da Costa, Palmira. "The Making of Extraordinary Facts: Authentication of Singularities of Nature at the Royal Society of London in the First Half of the Eighteenth Century." *Studies in history and philosophy of science* 33 2 (2002): 265–88.

———. *The Singular and the Making of Knowledge at the Royal Society of London in the Eighteenth Century*. Newcastle Upon Tyne: Cambridge Scholars Publishing, 2009.

Ford, John M.T. "A Medical Student at St Thomas's Hospital, 1801–1802. The Weekes Family Letters." Medical History, Supplement No. 7. London: Wellcome Institute for the History of Medicine, 1987.

Fuchs, Daniel, and Geo Fuchs. *Conserving*. Munich: Reuss, 2000.

Gaukroger, Stephen. *The Collapse of Mechanism and the Rise of Sensibility. Science and the Shaping of Modernity, 1680–1760*. Oxford: Oxford University Press, 2010.

Geerars, C.M. "Theorie Van De Satire." *Documentatieblad Werkgroep 18e eeuw* 15/16 (1971): 1–41.

Gettens, Rutherford J., Robert L. Feller, and W.T. Chase. "Vermilion and Cinnabar." *Studies in Conservation* 17.2 (1972): 45–69.

Geyskens, D.C., and C.J. van der Klaauw. "Der Heutige Zustand Der Anatomischen Kabinette Früherer Jahrhunderte in Leiden." *Janus Archives internationales pour l'histoire de la médicine et la géographie médicinale* 38 (1934): 179–92.

Goodwin, Charles. "Professional Vision." *American Anthropologist* 96 3 (1994): 606–33.

Gould, Stephen Jay, and Rosamond W. Purcell. *Finders, Keepers: Eight Collectors*. London: W.W. Norton, 1992.

Gowing, Laura. "Marked Bodies and Social Meanings." in *A Cultural History of the Human Body in the Elightenment*. Ed. Reeves, Carole. Vol. 4. Oxford/NY: Berg, 2010: 133–53.

Graham, Gordon. *Philosophy of the Arts. An Introduction to Aesthetics*. 2nd ed. London and New York: Routledge, 1997.

Greer, Germaine. "Gunther Von Hagens' Body Worlds Is Strictly Showbiz." *The Guardian* Monday February 25 2008.

Groningen, Museumhuis. "Universiteitsmuseum Groningen." *Extra Intermuseaal*. January (2011): 4.

Guépin, J.P. "Het Bataafse Oor: Drie Eeuwen Latijnse Poëzie Door Nederlanders Geschreven." *De Tweede Ronde: tijdschrift voor literatuur* 7.4 (1986): 135–48.

Guerrini, Anita. "Anatomists and Entrepreneurs in Early Eighteenth-Century London." *Journal of the History of Medicine and Allied Sciences* 59 2 (2004): 219–39.

———. "Duverney's Skeletons." *Isis* 94 4 (2003): 577–603.

———. *Experimenting with Humans and Animals: From Galen to Animal Rights. Johns Hopkins Introductory Studies in the History of Science*. Eds. Green, M.T. and S. Kingsland. Baltimore, Maryland: Johns Hopkins University Press, 2003.

———. "The Value of a Dead Body." in *Vital Matters. Eighteenth-Century Views of Conception, Life, and Death*. Eds. Deutsch, Helen and Mary Terrall. Toronto: University of Toronto Press, 2012: 246–264.

Guyer, Paul. "The difficulty of the sublime." *Histories of the Sublime 9–10 December 2004* Madelein, C., J. Pieters and B. Vandenabeele eds. Koninklijke Vlaamse Academie van Belgie voor Wetenschappen en Kunsten (2005): 33–43.

Gwilt, Joseph. "Aesthetic." *An encyclopædia of architecture, historical, theoretical, and practical*. Ed. Papworth, Wyatt. New York: Bonanza Books, 1842 (1982).

Hacking, Ian. *Historical Ontology*. Cambridge MA: Harvard University Press, 2002.

Hagner, Michael. "Enlightened Monsters" in *The Sciences in Enlightened Europe*. Eds. Clark, William, Jan Golinski and Simon Schaffer. Chicago & London: The University of Chicago Press, 1999: 175–217.

Hall, James. *Dictionary of Subjects and Symbols in Art*. 2nd ed. Boulder, CO: Westview Press, 2008.

Hallam, Elizabeth, and J. Hockey. "Death, Memory and Material Culture." in *Materializing Culture*. Eds. Gilroy, Paul, Michael Herzfeld and Danny Miller. Oxford: Berg, 2001.

———. "Articulating bones: an epilogue." *Journal of Material Culture* 15 4 (2010): 465–92.

Hansen, Julie. "Resurrecting Death: Anatomical Art in the Cabinet of Dr. Frederik Ruysch." *The art bulletin: an illustrated quarterly* 78 4 (1996): 663–79.

Hammermeister, Kai. *The German Aesthetic Tradition*. Cambridge: Cambridge University Press, 2002.

Harman, Graham. "Tool-Being: Elements in a Theory of Objects." Dissertation. DePaul University, Chicago, Ill. 1999.

Harvey, Karen. "Introduction: Practical Matters." in *History and Material Culture. A Student's Guide to Approaching Alternative Sources*. Ed. Harvey, Karen. London: Routledge, 2009: 1–23.

Helm, Jürgen, and Karin Stukenbrock, eds. *Anatomie. Sektionen Einer Medizinischen Wissenschaft Im 18. Jahrhundert*. Stuttgart: Franz Steiner Verlag, 2003.

Hemsterhuis, Frans. *Wijsgerige Werken. Uitgegeven, Ingeleid En Van Een Commentaar Voorzien Door M.J. Petry*. Budel / Leeuwarden: Damon / Fryske Akademy, 2001.

Hendriksen, Marieke. "Aesthesis in Anatomy. Materiality and Elegance in the Eighteenth-Century Leiden Anatomical Collections." Dissertation. Leiden University, 2012.

———. "Une Recherche Commune De « L'homo Perfectus ? La Perfection Dans La Relation De Travail De Bernard Siegfried Albinus (1697–1770) Et Jan Wandelaar (1692–1759)." *Alliage* 67 (2010): 22–34.

———. Hieke Huistra and Rina Knoeff. "Recycling Anatomical Preparations" in S. Alberti and E. Hallam (eds.) *Medical Museums*, London: Royal College of Surgeons of England, 2013: 74–87.

———. "The Fate of the Beaded Babies: Forgotten Early Colonial Anatomy" in R. Knoeff and R. Zwijnenberg (eds.) *The Fate of Anatomical Collections*, London: Ashgate, 2014.

Henschen, Folke, and Kenneth P. Oakley. *The Human Skull. A Cultural History*. London: Thames and Hudson, 1966.

Herman, Bernard L. "Introduction. The Discourse of Objects." in *The Stolen House*. Charlottesville: The University Press of Virginia, 1992. pp. 3–14.

Huisman, Tim & A.M. Luyendijk-Elshout. *De Volmaakte Mens: De Anatomische Atlas Van Albinus En Wandelaar. Mededeling Uit Het Rijksmuseum Voor De Geschiedenis Van De Natuurwetenschappen*. Vol. 0926–8936; 248. Leiden: Museum Boerhaave (Leiden), 1991.

———. *The Finger of God. Anatomical Practice in 17th Century Leiden*. Leiden: Primavera Press, 2009.

Huistra, Hieke. "Collecties Op College. Het Gebruik Van Anatomische Preparaten in Het Negentiende-Eeuwse Geneeskunde Onderwijs Aan De Nederlandse Universiteiten." in *Van Lectio Tot Powerpoint. De Geschiedenis Van Het Onderwijs Aan De Nederlandse Universiteiten*. Eds. Dorsman, L.J. and P.J. Knegtmans. Hilversum: Verloren, 2011: 25–42.

———. "Dead Body in the Closet: The Leiden Anatomical Collections, Their Meanings and Their Audiences in the Nineteenth Century." Dissertation, Leiden University, 2013.

Hyrtl, Joseph. *Handboek Der Praktische Ontleedkunst: Als Leiddraad Bij De Anatomische Oefeningen En Tot Het Vervaardigen Van Præparaten / Joseph Hyrtl; Naar Het Hoogduitsch Door J.A. Boogaard En T. Zaayer*. Trans. Boogaard, J.A. and T. Zaaijer. Schiedam, 1865.

Israel, Jonathan. *The Dutch Republic: Its Rise, Greatness and Fall, 1477–1806*. Oxford: Oxford University Press, 1995.

Jacob, Margaret, C., and W.W. Mijnhardt (eds). *The Dutch Republic in the eighteenth century: decline, Enlightenment, and revolution*. Ithaca: Cornell University Press, 1992.

Jardine, Nicholas. *The Scenes of Inquiry: On the Reality of Questions in the Sciences*. Oxford: Clarendon Press, 1991.

———. "Uses and Abuses of Anachronism in the History of the Sciences." *History of science* 38.121 (2000): 251–70.

Jorink, Eric. " 'Het Boeck Der Natuere': Nederlandse Geleerden En De Wonderen Van Gods Schepping, 1575–1715." Dissertation, Rijksuniversiteit Groningen, 2004.

———. *Reading the book of nature in the Dutch Golden Age, 1575–1715*. Leiden: Brill, 2010.

———. and Bart Ramakers. "Undivided territory. 'Art' and 'science' in the early modern Netherlands." in *Nederlands Kunsthistorisch Jaarboek* 61, Zwolle: WBooks, 2011.

Jorissen, W.P. *Het Chemisch (Thans Anorganisch Chemisch) Laboratorium Der Universiteit Te Leiden Van 1859–1909 En De Chemische Laboratoria Dier Universiteit Vóór Dat Tijdvak En Hen, Die Er in Doceerden*. Leiden: A.W. Sijthoff's Uitgevers-Maatschappij, 1909.

Jütte, Robert. "Die Haut Als Heilmittel." in *Verborgen Im Buch—Verborgen Im Körper: Haut Zwischen 1500 Und 1800*. Ed. Zeuch, Ulrike: Wolfenbüttel, 2003: 161–66.

Kavey, Allison B. "Mercury Falling. Gender Malleability and Sexual Fluidity in Early Modern Popular Alchemy." in *Chymists and Chymistry. Studies in the History of Alchemy and Early Modern Chemistry*. Ed. Principe, Lawrence M. Sagamore Beach: Chemical Heritage Foundation & Science History Publications, 2007: 125–35.

Kemp, Martin. "'The Mark of Truth': Looking and Learning in Some Anatomical Illustrations from the Renaissance and the Eighteenth Century." in *Medicine and the Five Senses*. Eds. Bynum, W.F. and Roy Porter. Cambridge: Cambridge University Press, 1993: 85–122.

———. and Marina Wallace. *Spectacular Bodies: The Art and Science of the Human Body from Leonardo to Now*. Hayward Gallery, 2000.

Klein, Ursula, and E.C. Spary. "Introduction: Why Materials?" in *Materials and Expertise in Early Modern Europe: Between Market and Laboratory*. Eds. Klein, Ursula and E.C. Spary. Chicago: University Of Chicago Press, 2010.

Kluveld, Amanda. *Reis door de hel der onschuldigen: de expressieve politiek van de Nederlandse anti-vivisectionisten, 1890–1940*. Amsterdam: Amsterdam University Press, 2000.

Knoeff, Rina. "Sex in Public. On the Spectacle of Female Anatomy in Amsterdam around 1700." *L'Homme. Europäische Zeitschrift für Feministische Geschichtswissenschaft*. 23.1 (2012): 43–58.

———. *Herman Boerhaave (1668–1738): Calvinist Chemist and Physician. Amsterdam*: Edita, 2002.

Knox, F. *The Anatomist's Instructor, and Museum Companion Being Practical Directions for the Formation and Subsequent Management of Anatomical Museums*. Edinburgh: Adam and Charles Black, 1836.

Konert, Jürgen, and Holger G. Dietrich. "Giovanni Battista Morgagni Und Der Beginn Der Pathologischen Anatomie" in *Anatomie. Sektionen Einer Medizinischen Wissenschaft Im 18. Jahrhundert*. Eds. Helm, Jürgen and Karin Stukenbrock. Stuttgart: Franz Steiner Verlag, 2003: 127–38.

Kooijmans, Luuc. *De Doodskunstenaar. De Anatomische Lessen Van Frederik Ruysch*. Amsterdam: Uitgeverij Bert Bakker, 2004.

———. Translated by Diane Webb. *Death Defied. The Anatomy Lessons of Frederik Ruysch*. History of Science and Medicine Library. Vol. 1872–0684. Leiden: Brill, 2009.

Korsmeyer, Carolyn. "Fear and Disgust: The Sublime and the Sublate." *Revue internationale de philosophie* 4.246 (2008): 367–79.

———. *Savoring Disgust. The Foul & the Fair in Aesthetics*. Oxford Oxford University Press, 2011.

Koudounaris, Paul. *The Empire of Death. A Cultural History of Ossuaries and Charnel Houses*. London: Thames & Hudson, 2011.

Kraatz, Anne. *Lace. History and Fashion*. London: Thames & Hudson, 1989.

La Mettrie, Julien Offray de. *Man a Machine. French-English Including Frederick the Great's Eulogy on La Mettrie and Extracts from La Mettries the Natural History of the Soul*. La Salle, Illinois: The Open Court Publishing Company, 1943.

Latour, Bruno. *Science in Action*. Cambridge, Massachusetts: Harvard University Press, 1987.

———. *We Have Never Been Modern*. Boston, MA: Harvard University Press, 1993.

———. *Reassembling the Social. An Introduction to Actor-Network Theory*. Clarendon Lectures in Management Studies. 2007 paperback edition ed. Oxford: Oxford University Press, 2005.

Laqueur, Thomas. *Making Sex: Body and Gender from the Greeks to Freud*. Cambridge, MA: Harvard University Press, 1992.

Le Fanu, W.R. "More Letters from B.S. Albinus to Robert Nesbitt." *Janus Archives internationales pour l'histoire de la médicine et la géographie médicinale* 36 (1932): 1–26.

Legêne, Susan. "Flatirons and the folds of history. On archives, cultural heritage and colonial legacies." in *Traveling Heritages. New Perspectives on Collecting, Preserving and Sharing Women's History*. Ed. S. Wierings. Amsterdam: Aksant, 2008: 47–64.

Lehmann, Ann-Sophie, and Herman Roodenburg. "Introduction." in *Netherlands Yearbook for Art History*. Eds. Lehmann, Ann-Sophie and Herman Roodenburg. Zwolle: Waanders Publishers, 2008: 7–13.

Leids Universitair Medisch Centrum (LUMC). "Catalogus Historische Anatomische Collectie LUMC." Ed. Leids Universitair Medisch Centrum. Leiden 2010. (Partly accessible via http://catalogue.leidenuniv.nl)

Leonard, C.H. and Gray, Henry. *The Concise Gray's Anatomy*. New York: Cosimo, 2005.

Leopardi, Giacomo. "Dialogi Di Frederico Ruysch E Delle Sue Mummie." in *Operette Morali*. Milaan: A.F. Stella, 1827: 163–71.

Lewis, Stephen. "A Pathological Misunderstanding." *Wellcome History Spring* 46 (2011): 6–7.

Lindeboom, G.A. (ed) *Observationes anatomicæ Collegii privati Amstelodamensis. Pars prior (1667) et altera (1673). Fascimile with an introduction by G.A. Lindeboom*. Nieuwkoop: B. De Graaf, 1975.

———. "Inleiding Bij De Surinaamsche Artz." in *Uitgaven Van De Natuurwetenschappelijke Studiekring Voor Suriname En De Antillen. Natuurhistorische Reeks Vol. 109*. Amsterdam: Rodopi, 1981.

Luck, Georg. "Love Elegy." *Latin Literature*. Eds. Kenney, E.J. and W.V. Clausen. Cambridge: Cambridge University Press, 1982.

LUMC catalogue: see Leids Universitair Medisch Centrum.

Luyendijk-Elshout, A.M. "Antony Nuck (1650–1692) the "Mercator" of the Body Fluids. A Review of His Anatomical and Experimental Studies." in *Circa Tiliam. Studia Historiae Medicinae Gerrit Arie Lindeboom Septuagenario Oblata*. Eds. Meer, C. van der, J.V. Meininger and J. Schouten. Leiden: E.J. Brill, 1974: 150–64.

———. "Death Enlightened: A Study of Frederik Ruysch." *Journal of the American Medical Association* 212 (1970): 121–6.

———. "Introduction." in *Dilucidatio Valvularum in Vasis Lymphaticis Et Lacteis, 1665 / Frederik Ruysch. Facs. Ed. / with an Introd. By A.M. Luyendijk-Elshout*. Nieuwkoop: De Graaf, 1964.

———. "Morgagni, Ruysch, E Boerhaave. I Tre Moschettieri Dell'anatomia Sottile." *Scienza E Cultura, Informazione Dell'università Di Padova*. Padova: Università di Padova, 1983: 301–20.

———. "The Elegant Anatomist: The Italian Medical Connections of Eduard Sandifort (1742–1814)." in *Italian Scientists in the Low Countries in the xviith and xviiith Centuries*. Eds. Maffioli, C.S. and L.C. Palm. Nieuwe Nederlandse Bijdragen Tot De Geschiedenis Der Geneeskunde En Der Natuurwetenschappen. Amsterdam: Rodopi, 1989: 305–20.

Maerker, Anna. "Model Experts: The Production and Uses of Anatomical Models at La Specola, Florence, and the Josephinum, Vienna, 1775–1814." Dissertation, Cornell University, 2005.

———. *Model experts: wax anatomies and Enlightenment in Florence and Vienna, 1775–1815*. Machester: Manchester University Press 2011.

Malik, Kenan. *The Meaning of Race: Race, History and Culture in Western Society*. Macmillan: Basingstoke, 1996.

Margócsy, Dániel. "Advertising Cadavers in the Republic of Letters: Anatomical Publications in the Early Modern Netherlands." *British Journal for the History of Science* 42 2 (2008): 187–210.

———. "A Museum of Wonders or a Cemetery of Corpses? The Commercial Exchange of Anatomical Collections in Early Modern Netherlands." in *Silent Messengers. The Circulation of Material Objects of Knowledge in the Early Modern Low Countries*. Eds. Dupré, Sven and Cristoph Lüthy. Berlin: Lit Verlag, 2011: 185–216.

Mazzolini, Renato G. "Las Castas: Interracial Crossing and Social Structure, 1770–1835." in *Heredity Produced. At the Crossroads of Biology, Politics, and Culture, 1500–1870* Eds. Müller-Wille, Staffan and Hans-Jörg Rheinberger. Transformations: Studies in the History of Science and Technology. Cambridge, MA/London, England: The MIT Press, 2007: 349–73.

McClellan, James E. *Science Reorganized. Scientific Societies in the Eighteenth Century.* New York: Columbia University Press, 1985.

Meerman, Johan. *Eenige Berichten Omtrent Het Noorden En Noord-Oosten Van Europa.* 4 vols. Den Haag: Erven Js. van Cleef, 1804.

Menninghaus, Winfried. *Disgust: The Theory and History of a Strong Sensation.* Trans. EIland, Howard and Joel Golb. Albany: State University of New York Press, 2003.

Messbarger, Rebecca. *The Lady Anatomist. The Life and Work of Anna Morandi Manzolini.* Chicago and London: The University of Chicago Press, 2010.

Meyer, Birgit. *Religious Sensations. Why Media, Aesthetics and Power Matter in the Study of Contemporary Religion.* Amsterdam: University of Amsterdam, 2006.

Miller, William Ian. *The Anatomy of Disgust.* Cambridge, MA: Harvard University Press, 1997.

Molhuysen, P.C. *Bronnen Tot De Geschiedenis Der Leidsche Universiteit. Rijks Geschiedkundige Publicatiën.* 7 Vols. 's Gravenhage: Martinus Nijhoff, 1913–1924. (Accessible via: http://www.historici.nl/retroboeken/leidsche_universiteit/#page=0&size=800&accessor=toc&source=1)

Montagu, Lady Mary Wortley "Saturday; Small-Pox." in *The Letters and Works of Lady Mary Wortley Montagu. Ed. Wharncliffe, Lord.* Vol. 3. London: Richard Bentley, 1837: 366–9.

Moore, Wendy. *The Knife Man. The Extraordinary Life and Times of John Hunter, Father of Modern Surgery.* London: Bantam Press, 2005.

Moran, Bruce T. "Introduction." *Isis* 102.2 (2011): 300–04.

Morton, Timothy. *Ecology Without Nature: Rethinking Environmental Aesthetics.* Cambridge, MA: Harvard University Press, 2007.

Mueller-Wille, Staffan, and Hans-Jörg Rheinberger. "Heredity—the Formation of an Epistemic Space." in *Heredity Produced. At the Crossroads of Biology, Politics, and Culture, 1500–1870.* Eds. Müller-Wille, Staffan and Hans-Jörg Rheinberger. Transformations: Studies in the History of Science and Technology. Cambridge, MA/London, England: The MIT Press, 2007: 3–34.

Newman, William R. *Atoms and Alchemy. Chymistry and the Experimental Origins of the Scientific Revolution.* Chicago: The University of Chicago Press, 2006.

———. "From Alchemy to 'Chymistry'." in *The Cambridge History of Science.* Eds. Park, Katharine and Lorraine Daston. *Vol. 3 Early Modern Science.* Cambridge: Cambridge University Press, 2006.

———. and Lawrence M. Principe. "Alchemy Vs. Chemistry: The Etymological Origins of a Historiographic Mistake." *Early Science and Medicine* 3.1 (1998): 32–65.

Nisbet, R.G.M., and Margaret Hubbard. *A Commentary on Horace: Odes, Book I.* Oxford: Oxford University Press, 2001.

Noble, Louise Christine. " 'And Make Two Pasties of Your Shameful Heads': Medicinal Cannibalism and Healing the Body Politic in Titus Andronicus." *ELH* 70 3 (2003): 677–708.

Nutton, Vivian. "The fatal embrace; Galen and the history of ancient medicine", *Science in Context*, 2005, 18, 1: 111–21.

———. "The fortunes of Galen", in R.J. Hankinson, ed., *The Cambridge Companion to Galen*, Cambridge, Cambridge University Press, 2008, 355–90.

Ololade, Olatunji. "Where Body Parts Are Called Credit Cards." The Nation 30/07/2011 2011, Online ed. http://www.thenationonlineng.net/2011/index.php/columnist/friday/olatunji-ololade/14038-where-body-parts-are-called-credit-cards.html.

Olry, Regis, and Kaoru Motomiya. "Paolo Mascagni, Ernest Alexandra Lauth and Marie Philibert Constant Sappey on the Dissection and Injection of the Lymphatics." *Journal of the International Society for Plastination* 12.2 (1997): 4–7.

Onnasch, Ernst-Otto. "Receptieonderzoek. De Eerste Receptie Van Kants Filosofie in Nederland." *Tijdschrift voor filosofie* 68 (2006): 133–56.

Otterspeer, Willem. *Groepsportret Met Dame II. De Vesting Van Macht. De Leidse Universiteit, 1673–1775*. Amsterdam: Uitgeverij Bert Bakker, 2002.

———. *Groepsportret Met Dame III. De Werken Van Wetenschap 1776–1876*. Amsterdam: Bert Bakker, 2005.

Park, Katherine. *Secrets of Women: Gender, Generation, and the Origins of Human Dissection*. New York: Zone Books, 2006.

Pemberton III, John. *African Beaded Art: Power and Adornment*. Northampton, MA: Smith College Museum of Art, 2008.

Pickstone, John V. *Ways of Knowing: A New History of Science, Technology and Medicine*. Chicago: University of Chicago Press, 2001.

Pinney, Christopher. "Four Types of Visual Culture." in *Handbook of Material Culture*. Ed. Tilley, C. London: Sage, 2006: 131–45.

Pochat, Götz. *Geschichte der Ästhetik und Kunsttheorie. Von der Antike bis zum 19. Jahrundert*. Köln: DuMont Buchverlag, 1986.

Pol, Arent. *Schepen Met Geld. De Handelsmunten Van De Verenigde Oostindische Compagnie 1602–1799*. Den Haag: SDU Uitgeverij, 1989.

Porter, Roy. *Flesh in the Age of Reason*. London: Allen Lane, Penguin Books, 2003.

———. (ed.) *Cambridge Illustrated History of Medicine*. Cambridge: Cambridge University Press, 2009.

Powers, John C. "Scrutinizing the Alchemists. Herman Boerhaave and the Testing of Chymistry." in *Chymists and Chymistry. Studies in the History of Alchemy and Early Modern Chemistry*. Ed. Principe, Lawrence M. Sagamore Beach: Chemical Heritage Foundation & Science History Publications, 2007: 227–38.

———. *Inventing Chemistry. Herman Boerhaave and the Reform of the Chemical Arts*. Chicago/London: University of Chicago Press, 2012.

Priester, L.R. *De Nederlandse Houding Ten Aanzien Van De Slavenhandel En Slavernij, 1596–1863. Het Gedrag Van De Slavenhandelaren Van De Commercie Compagnie Van Middelburg in De 18e Eeuw*. Scriptiereeks Crgz. Middelburg: Commissie Regionale Geschiedbeoefening Zeeland, 1987.

Principe, Lawrence M. "Alchemy Restored." *Isis* 102.2 (2011): 305–12.

Prown, Jules David. "Mind in Matter: An Introduction to Material Culture Theory and Method." *Winterthur Portfolio* 17.1 (1982): 1–19.

———. *Art as Evidence. Writings on Art and Material Culture*. New Haven/London: Yale University Press, 2001.

Punt, H. "Bernard Siegfried Albinus (1697–1770). On 'Human Nature'. Anatomical and Physiological Ideas in Eighteenth Century Leiden." Dissertation, Leiden University, 1983.

Rheinberger, Hans-Jörg. *On Historicizing Epistemology: An Essay. Cultural Memory in the Present*. Stanford, CA: Stanford University Press, 2010.

Ricardo Brown, B. *Until Darwin. Science, Human Variety and the Origins of Race*. London: Pickering & Chatto, 2010.

Richardson, Ruth. *Death, Dissection, and the Destitute*. Chicago: University of Chicago Press, 2001.

———. *The Making of Mr. Gray's Anatomy*. Oxford: Oxford University Press, 2008.

Riskin, Jessica. *Science in the Age of Sensibility: The Sentimental Empiricists of the French Enlightenment*. Chicago: University of Chicago Press, 2002.

Roberts, Lissa. "Centres and Cycles of Accumulation." in *Centres and Cycles of Accumulation in and around the Netherlands During the Early Modern Period*. Lissa Roberts (ed). Vienna: Lit Verlag, 2011: 3–27.

———. "The Death of the Sensuous Chemist: The 'New' Chemistry "And the Transformation of Sensuous Technology." in *Empire of the Senses. The Sensual Culture Reader*. Ed. Howes, David. Oxford: Berg, 2005: 106–27.

———. Simon Schaffer, and Peter Dear, eds. *The Mindful Hand. Inquiry and Invention from the Late Renaissance to Early Industrialisation*. Amsterdam: Koninklijke Nederlandse Akademie van Wetenschappen, 2007.

Ronson, Albert R. *The new medicine and the old ethics*. Cambridge MA: Harvard University Press, 1990.

Rookmaaker, L.C., and R.P.W. Visser. "Petrus Camper's Study of the Javan Rhinoceros (Rhinoceros Sondaicus) and Its Influence on Georges Cuvier." *Bijdragen tot de Dierkunde* 52 2 (1982): 121–36.

Rosner, Lisa. *The Anatomy Murders: Being the True and Spectacular History of Edinburgh's Notorious Burke and Hare, and of the Man of Science Who Abetted Them in the Commission of Their Most Heinous Crimes*. Philadelphia: University of Pennsylvania Press, 2010.

Rouffaer, G.P. "Waar Kwamen De Raadselachtige Moetisalah's (Aggri-Kralen) in De Timor-Groep Oorspronkelijk Van Daan?" *Bijdragen tot de taal-, land- en volkenkunde van Nederlandsch-Indië, uitgegeven door het Koninklijk Instituut voor de taal-, land- en volkenkunde van Nederlandsch-Indië VI* (1899): 409–675.

Rousseau, G.S. *Nervous Acts. Essays on Literature, Culture and Sensibility*. Basingstoke / New York: Palgrave Macmillan, 2004.

Rutten, A.M.G. *Dutch Transatlantic Medicine in the Eighteenth Century under the Cover of the West India Company*. second ed. Rotterdam: Erasmus Publishing, 2000.

Rutten, Willibrord. "'De Vreselijkste Aller Harpijen': Pokkenepidemieën En Pokkenbestrijding in Nederland in De Achttiende En Negentiende Eeuw: Een Sociaal-Historische En Historisch-Demografische Studie." Dissertation, Landbouwuniversiteit Wageningen, 1997.

Sacré, Dirk. "An Imitator of Fracastorius's Syphilis: Gadso Coopmans (1746–1810) and His Varis." *Humanistica Lovaniensia* 45 (1996): 520–38.

Sandifort, G. *Museum Anatomicum Academiae Lugduno-Batavae Descriptum*. Plano with test prints vols. Lugd. Bat, 1827.

Santing, Catrien. "Andreas Vesalius's De Fabrica corporis humana, depiction of the human model in word and image." in *Body and Embodiment in Netherlandish Art/ Lichaam en lichamelijkheid in de nederlandse kunst. Netherlands Yearbook for Art History*. Eds. Lehmann, Ann-Sophie and Herman Roodenburg. Zwolle: Waanders Publishers, 2008: 59–85.

———. and Jetze J. Touber (eds). *Blood-Symbol-Liquid*. Groningen Studies in Cultural Change (H.W. Hoen, ed.), vol. XLIV. Leuven: Peeters, 2012.

———. "De Menselijke Canon. Het Vraagstuk Van De Ideale Mens in De Nederlandse Medische Wetenschap." *Bijdragen en mededelingen betreffende de geschiedenis der Nederlanden / The Low Countries historical review* 122 1 (2007): 484–503.

Schaffer, Simon. "Genius in Romantic Natural Philosophy." in *Romanticism and the Sciences*. Eds. Cunningham, Andrew and Nicholas Jardine. Cambrdidge: Cambridge University Press, 1990: 82–98.

Schickore, Jutta. *The Microscope and the Eye: A History of Reflections, 1740–1870*. Chicago and London: The University of Chicago Press, 2007.

Schiebinger, Londa. *Plants and Empire. Colonial Bioprospecting in the Atlantic World*. Cambridge MA/London UK: Harvard University Press, 2004.

Schlothauer, Andreas. "Kopfjagd Und Schädelkult in West-Afrika." in *Schädelkult. Kopf Und Schädel in Der Kulturgeschichte Des Menschen. Begleitband Zur Sonderausstellung. Ed. Wieczorek, Alfried. Reiss-Engelhorn-Museen*. Mannheim: Schnell Steiner, 2012: 117–23.

Schmidt, Benjamin. "Accumulating the World: Collecting and Commodifying 'Globalism' in Early Modern Europe." in *Centres and Cycles of Accumulation in and around the Netherlands During the Early Modern Period.* Lissa Roberts (ed). Vienna: Lit Verlag, 2011: 129–154.

Schultka, Rüdiger, and Luminita Göbbel. "Präparationstechniken Und Präparate Im 18. Und Frühen 19. Jahrhundert, Dargestellt an Beispielen Aus Den Anatomischen Sammlungen Zu Halle (Saale)." in *Anatomie. Sektionen Einer Medizinischen Wissenschaft Im 18. Jahrhundert.* Eds. Helm, Jürgen and Karin Stukenbrock. Stuttgart: Franz Steiner Verlag, 2003: 49–82.

Shapin, Steven, and Simon Schaffer. *Leviathan and the Air-Pump. Hobbes, Boyle, and the Experimental Life.* Princeton, NJ: Princeton University Press, 1985.

———. *Never Pure. Historical Studies of Science as If It Was Produced by People with Bodies, Situated in Time, Space, Culture, and Society, and Struggling for Credibility and Authority.* Baltimore: The John Hopkins University Press, 2010.

Sjtsjedrova, Irina. "Abraham Kaau-Boerhaave: bladzijden uit de biografie van een academicus". In *Noord- en Zuid-Nederlanders in Rusland 1703–2003,* bewerkt door E. Waegemans, J.S.A.M. Koningsbrugge, en Nadja Louwerse, 293–312. Baltic studies 9. Groningen: INOS, 2004.

Smith, Pamela H. *The Body of the Artisan. Art and Experience in the Scientific Revolution.* Chicago: University of Chicago Press, 2004.

———. 'The Workshop of History. Reconstructing Lived Experience in Early Modern Europe' Huygens-Descartes lecture 2011, Amsterdam, 15 June 2011 (unpublished).

———. "Vermillion, Mercury, Blood and Lizards: Matter and Meaning in Metalworking." in *Materials and Expertise in Early Modern Europe: Between Market and Laboratory.* Eds. Klein, Ursula and E.C. Spary. Chicago: University Of Chicago Press, 2010: 29–49.

———. and Benjamin Schmidt, eds. *Making Knowledge in Early Modern Europe: Practices, Objects, and Texts, 1400–1800.* Chicago & London: The University of Chicago Press, 2007.

———. and Tonny Beentjes. "Nature and Art: Making and Knowing: Reconstructing Sixteenth-Century Life-Casting Technniques." *Renaissance Quarterly* 63 1 (2010): 128–79.

Snelders, H.A.M. "Het Bataviaasch Genootschap Van Kunsten En Wetenschappen in De Periode 1778 Tot 1816." *Documentatieblad werkgroep Achttiende eeuw.* 41/42 (1979): 62–90.

Snelders, Stephen. *Vrijbuiters van de heelkunde. Op zoek naar medische kennis in de tropen 1600–1800.* Amsterdam/Antwerpen: Uitgeverij Atlas, 2012.

Spacks, Patricia Meyer. *Reading Eighteenth-Century Poetry. Reading Poetry.* Chichester: Wiley-Blackwell, 2009.

Spary, E.C. "Of Nutmegs and Botanists: The Colonial Cultivation of Botanical Identity." in *Colonial Botany. Science, Commerce, and Politics in the Early Modern World.* Eds. Schiebinger, Londa and Claudia Swan. Philadelphia: University of Pennsylvania Press, 2005: 187–203.

Spillane, John D. *The Doctrine of the Nerves. Chapters in the History of Neurology*. Oxford: Oxford University Press, 1981.

Staves, Susan. "The Puzzle of the Pox-Marked Body." in *A Cultural History of the Human Body in the Elightenment*. Ed. Reeves, Carole. Vol. 4. Oxford/NY: Berg, 2010: 155–74.

Steinke, Hubert. *Irritating Experiments. Haller's Concept and the European Controversy on Irritability and Sensibility, 1750–90*. The Wellcome Series in the History of Medicine. Eds. V. Nutton, C.J. Lawrence and M. Neve. Amsterdam-New York: Rodopi, 2005.

Stoll, Mareidi, and Gert Stoll. *Ibeji. Zwillingsfiguren Der Yoruba / Twin Figures of the Yoruba*. Düsseldorf: Hub. Hoch, 1980.

Sturm, Lars-Burkhardt. "Präparationstechniken Und Ihre Anwendung in Den Meckelschen Sammlungen Zu Halle/Saale." in *Anatomie Und Anatomische Sammlungen Im 18. Jahrhundert. Anlässlich Der 250. Widerkehr Des Geburtstages Von Philip Friedrich Theodor Meckel (1755–1803)*. Eds. Schultka, Rüdiger and Josef N. Neumann. Berlin: Lit Verlag, 2007: 377–88.

Sugg, Richard. *Mummies, Cannibals And Vampires. The History Of Corpse Medicine From The Renaissance to the Victorians*. London: Routledge, 2011.

Suringar, G.C.B. "Het Bijeenbrengen Eener Verzameling Van Natuurlijke Voorwerpen Voor Het Akademisch Onderwijs, Omstreeks Het Midden Der Achttiende Eeuw. Eerste Afzonderlijke Lessen over De Natuurlijke Historie, Door Den Hoogleraar Allamand, En over De Zoölogie Door Le Francq Van Berkhey." *Ned Tijdschr Geneeskd*. 11 (1867): 265–84.

Swan, Claudia. "Collecting Naturalia in the Shadow of Early Modern Dutch Trade." in *Colonial Botany. Science, Commerce, and Politics in the Early Modern World*. Eds. Schiebinger, Londa and Claudia Swan. Philadelphia: University of Pennsylvania Press, 2005: 223–36.

Swiderski, Richard M. *Quicksilver: A History of the Use, Lore and Effects of Mercury*. Jefferson, NC: McFarland, 2008.

Thomas, Simon. "Teunis Zaaijer." *Nieuw Nederlandsch biografisch woordenboek*. Vol. 4. Eds. Molhuysen and P.J. Blok. Leiden: Sijthoff, 1918.

Tompsett, D. H. *Anatomical Techniques*. Edinburgh: E. & S. Livingstone, 1970.

Trivellato, Francesca. "Venetian Glass Beads, Female Labour and International Trades." in *Beads and Bead Makers. Gender, Material Culture and Meaning*. Eds. Scimia, Lidia D. and Joanne B. Eicher. Oxford: Berg, 1998: 47–82.

Van Delft, Dirk. "De Schoonheid En De Dood." NRC Handelsblad Zaterdag 18 december 2004.

Van der Borg, Els. "Het Ontslag Van De Stadsvroedmeester Cornelis Terne. Wedijverende Beroepsgroepen Te Leiden in De Achttiende Eeuw." in *Holland: regionaal-historisch tijdschrift* 22 (1990): 109–20.

Van der Korst, J.K. *Een Dokter Van Formaat: Gerard Van Swieten, Lijfarts Van Keizerin Maria Theresia*. Houten: Bohn Stafleu van Loghem, 2003.

Van de Roemer, Gijsbert M. "From Vanitas to Veneration. The Embellishments in the Anatomical Cabinet of Frederik Ruysch." *Journal of the History of Collections* (2009): 1–18.

Van der Zwaag, Pieter. "Wouter Van Doeveren (1730–1783): Leven En Werken Van Een 18e Eeuws Hoogleraar in De Geneeskunde." Dissertation, Amsterdam VU, 1970.

Van der Zwan, Nelleke. *Oog Voor Kralen*. Berg en Dal: Afrikamuseum, 1985.

Van Duuren, David, et al. *Physical Anthropology Reconsidered. Human Remains at the Tropenmuseum. Bulletins of the Royal Tropical Institute*. Ed. Kouwenhove, Arlette. Vol. 375. Amsterdam: KIT Publishers, 2007.

Van Gelder, Esther. *Tussen Hof En Keizerskroon. Carolus Clusius En De Ontwikkeling Van De Botanie Aan Midden-Europese Hoven (1573–1593)*. Leiden: Leiden University Press, 2011.

Van Heiningen, Teunis Willem. *Gerard Sandifort (1799–1848) in Twee Werelden*. Utrecht: Elinkwijk BV, 1995.

———. *The Correspondence of Sebald Justinus Brugmans (1763–1819)*. 2009, http://www.dwc.knaw.nl/archive/images/PDF/2009-vHeiningen-Correspondence%20Brugmans.pdf.

Verheij, Irene. *Op Reis Met Clara. De Geschiedenis Van Een Bezienswaardige Neushoorn*. Rotterdam: Natuurmuseum, 1992.

Vogelzang, Jacques. *Ere Ibeji. Beelden Van Tweelingen, in Het Bijzonder Van De Yoruba Van Nigeria*. Berg en Dal: Afrikamuseum, 1994.

Von Goethe, Johann Wolfgang. *Goethes Werke*. München: Deutscher Taschenbuch Verlag, 1987.

Von Haller, Albrecht. "Albrecht Hallers Tagebücher Seiner Reise Nach Holland" *Haller in Holland. Dagboek*. Ed. Lindeboom, G.A. Delft: Koninklijke Gist- en Spiritusfabriek N.V., 1725–1727 (1958).

Weber, Andreas. *Hybrid Ambitions. Science, governance, and empire in the career of Caspar G.C. Reinwardt (1773–1854)*. Leiden: Leiden University Press, 2012.

West Fitzhugh, Elisabeth. "Red Lead and Minimum." *Artists' Pigments: A Handbook of Their History and Characteristics*. Ed. Feller, Robert L. Vol. 1. London/Washington DC: Cambridge University Press/National Gallery of Art, 1986: 109–39.

Wieczorek, Alfried, ed. Schädelkult. *Kopf Und Schädel in Der Kulturgeschichte Des Menschen. Begleitband Zur Sonderausstellung*. Mannheim: Schnell Steiner, 2012.

Wilcox, A.J., et al. "Folic Acid Supplements and the Risk of Facial Clefts: A National Population-Based Control Study." *British Medical Journal* 334.7591 (2007).

Wilson, Philip K. "Eighteenth-Century 'Monsters' and Nineteenth-Century 'Freaks': Reading the Maternally Marked Child." *Literature and Medicine*, 21 1, (2002): 1–25.

Witkam, H.J. "Catalogues Anatomy Hall Leiden University." Leiden: Leiden University, 1980.

———. "Over De Anatomieplaats, De Albinussen En De Sandiforts." Leiden, 1968.

Zaaijer, T. *Beschrijving van twee vrouwenbekkens uit den Oost-Indischen Archipel.* Leiden: Doesburgh, 1862.

———. *Het Gewigt Eener Doelmatige Ontleedkundige Techniek. Redevoering Ter Aanvaarding Van Het Hoogleraarsambt Aan De Hoogeschool Te Leiden.* Leiden: Firma Jacs. Hazenberg, Corns. Zoon, 1866.

Zijlmans, Jori. Hond & Baas. *Een Geschiedenis Van Haat En Liefde.* Den Haag: De Nieuwe Haagsche / Haags Historisch Museum, 2002.

Zuidervaart, Huib J. "Egbert Philip Van Visvliet (1736–1799); Een Middelburgs Verzamelaar Van Vrienden, Naturalia En Fysische Instrumenten" *Zeeland* 3 2 (1994): 87–95.

Online Resources

Aesop. *Aesop's Fables: Complete, Original Translation from Greek. By Aesop. Translated Form the Greek by George Fyler Townsend (1887).* Ed. forgottenbooks.org. 2007. www.forgottenbooks.org.

Camper collection online: http://anatomie.ub.rug.nl/index.htm.

Geschiedeniszeeland.nl. "Kralen". 2009. http://www.geschiedeniszeeland.nl/themas/slavernij/handelsgoederen/kralen?lng=nl.

Groningen University Museum: http://www.rug.nl/museum/tentoonstellingen/actueel/BinnensteBuiten.

Historical calculator: http://www.iisg.nl/hpw/calculate.php.

Kunstkamera, St. Petersburg. "Frederik Ruysch's Collection." MAE RAS, 2010. http://www.kunstkamera.ru/kunst-catalogue/index.seam?c=RUYSH.

Leids Universitair Medisch Centrum (LUMC). "Anatomisch Museum Lumc". Leiden, 2012. Leids Universitair Medisch Centrum. https://www.lumc.nl/org/studentenportaal/voorzieningen/anatomisch-museum/.

Nijmegen University Anatomical Museum: http://www.umcn.nl/ONDERWIJS/MUSEUMVOORANATOMIEENPATHOLOGIE/Pages/default.aspx.

The Chymistry of Isaac Newton: http://webapp1.dlib.indiana.edu/newton/reference/chemLab.do.

"Voc Opvarenden". The Hague, 2008. Nationaal Archief. http://vocopvarenden.nationaalarchief.nl/detail.aspx?ID=1270029.

Von Haller, Albrecht. to Carl Linnaeus, 14 April 1737, The Linnaean correspondence, linnaeus.c18.net, letter L0182.

Index

www.ingramcontent.com/pod-product-compliance
Lightning Source LLC
LaVergne TN
LVHW082320171026
850499LV00004B/103
* 9 7 8 9 0 0 4 2 6 2 7 8 2 *